WHY ANIMALS MATTER

WHY ANIMALS MATTER

*ANIMAL CONSCIOUSNESS, ANIMAL
WELFARE, AND HUMAN WELL-BEING*

MARIAN STAMP DAWKINS

OXFORD
UNIVERSITY PRESS

OXFORD

UNIVERSITY PRESS

Oxford University Press, Inc., publishes works that further
Oxford University's objective of excellence
in research, scholarship, and education.

Oxford   New York
Auckland   Cape Town   Dar es Salaam   Hong Kong   Karachi
Kuala Lumpur   Madrid   Melbourne   Mexico City   Nairobi
New Delhi   Shanghai   Taipei   Toronto

With offices in
Argentina   Austria   Brazil   Chile   Czech   Republic   France   Greece
Guatemala   Hungary   Italy   Japan   Poland   Portugal   Singapore
South   Korea   Switzerland   Thailand   Turkey   Ukraine   Vietnam

Copyright © 2012 by Marian Stamp Dawkins

Oxford University Press, Inc.
198 Madison Avenue, New York, New York 10016

www.oup.com

Oxford is a registered trademark of Oxford University Press
Published in the UK by Oxford University Press Inc.

Library of Congress Cataloging-in-Publication Data
Dawkins, Marian Stamp.
Why animals matter : animal consciousness, animal welfare,
and human well-being / Marian Stamp Dawkins.
p.   cm.
Includes bibliographical references and index.
ISBN 978-0-19-974751-1
1. Animal welfare.   2. Consciousness in animals.
3. Human-animal relationships.   I. Title.
HV4708.D3824   2012
179'.3—dc23
2011051870

1 3 5 7 9 8 6 4 2

Printed in the United States of America
on acid-free paper

CONTENTS

PREFACE

I wrote this book because I am concerned that the wrong case for animal welfare is currently being put forward. There is too much reliance on anthropomorphism and not enough on science, too much emphasis on the conscious experiences animals may have and not enough on the value animals have for human well-being. As a result, animal welfare is losing out. It is failing to make use of some very powerful arguments that could be used to promote the welfare of animals. My aim is to correct this imbalance by stimulating a rethinking of attitudes to animals. I want to leave both them and us better off as a result.

The book is aimed as much at people who do not think that the welfare of non-human animals is particularly important as at those who are already convinced that it is. It makes a direct criticism of an anthropomorphic approach to animals and shows how both animals and humans benefit from a more scientific approach to animal welfare.

To avoid unnecessary words, I have used the word 'animal' to mean 'non-human animal' as opposed to 'human' throughout the book. Although humans are animals too, putting in the words 'non-human' every time becomes tedious and breaks the flow of the text. Where humans are included, this is stated explicitly.

The ideas in this book come from many sources and I would particularly like to acknowledge the importance of conversations with Roland Bonney, Temple Grandin, and Ruth Layton for showing me that good welfare for animals is not good welfare unless it can be made to work in practice, with Paul Cook and Malcolm Pye for their inspirational conviction that if something doesn't work the first time, it can always be made to work better next time, and with Edmund Rolls for essential clarifications about what consciousness and emotion are and, more importantly, what they are not. Latha Menon made some invaluable comments on the manuscript. I am also grateful to numerous other people for sharing with me their hopes, their concerns, their constraints, and sometimes even their data.

Marian Stamp Dawkins
September 2011

1

NO ROOM ON THE AGENDA

This is a book about animal welfare, but with a difference. It is not about how beautiful and amazing animals are (although they are), nor does it try to persuade you to change the way you treat them (although you may end up doing so). It is about something altogether more radical and, in its way, more disturbing. It is about the need to rethink our attitudes to the billions of non-human animals that inhabit our planet with us.

Why should we need to do any such rethinking? There are two reasons. The first is that the growing public concern with other issues such as climate change and how to feed the current and future human occupants of the earth is showing an alarming tendency to push animal welfare off the political agenda. The UN's Food and Agriculture Organization's publication, *Livestock's Long Shadow*,[1] for example, recommends the intensification of farming methods as a way of increasing food production and reducing greenhouse gases but does not discuss what this might mean for animal welfare in the agriculture of the future. The words 'animal welfare' appear only three times in the report's 390 pages and one of those is in a footnote. Similarly, the 2011 Foresight Report on the future of food and farming also refers to animal welfare in passing (five mentions in 206 pages) but does not put animal welfare anywhere in its five main priorities for farming.[2] And a recent high-profile review of how to feed the estimated 9 billion people there will be on earth by 2050 does not even mention animal welfare once. Food production for humans, reducing greenhouse gases, and maintaining biodiversity all take priority over the fate of non-humans.[3] At the conference tables and the seminars where the future of the planet is being discussed, there appears to be no seat for animal welfare because human health, economic costs, environmental impact have already taken most of the space and are speaking with much louder voices. Is this pragmatism or short-sightedness? Are we right to let human welfare take such priority over animal welfare, or are we overlooking the importance of animal health and welfare to the future of humans? Whatever your views about animals themselves, we all need answers to such questions for the security of our own futures.

The second reason is the extraordinary confusion that exists about non-human animals, particularly whether or not they have conscious experiences like us. For some people, it is obvious that they do and that therefore 'speciesism' (discriminating against other individuals because of which species they belong to) is on a level with racism or sexism.[4] For others, it is equally obvious that animals do not have conscious experiences, or that if they do, their experiences do not matter, at least not in comparison with any-thing a human might experience.[5] As a result, human attitudes to non-humans are a tangled mixture of sentiment, prejudice, self-interest, and practicality. Consistent they are not. Some animals are cherished as part of human families, others vilified as pests, and yet others viewed as though they were little more than inani-mate products of a factory. Attitudes vary from culture to culture and from individual to individual. Everywhere you look, there is confusion about so many animal-related issues—what animal consciousness is, which animals have emotions, whether animals can be said to have a 'quality of life', and whether we can ever agree on what good 'welfare' is are just a few of the main ones, but there are many others. From respected scientific journals to the popular press, from casual conversation to government legislation, there is confusion and misunderstanding. No wonder animal welfare is not making itself heard sufficiently.

These two problems with animal welfare—failing to get a look in when more pressing issues come along and confusion about basic ideas—are clearly connected. We are faced with the most extraordinary and unprecedented series of problems in the his-tory of our planet.[6] One species of ape—us—is so successful that it threatens the whole future of the earth as a habitable place. We are having to make decisions about what to eat, how to use scarce resources, what areas of the globe to preserve in their present state, and which to exploit. We are making decisions that could affect all life on earth and affect the future of many non-human species as well as our own. And yet, the ways in which we think about animals and the arguments that are put forward often lack coherence or proper evidence. Anyone who wants to be sure of keeping animal welfare on the political agenda in the future will need more coherent arguments and better evidence than are currently used. That means more science. It is a basic goal of this book to show that the best case for animal welfare comes from good science, both about the animals themselves and about their impact on us.

There is no getting away from the fact that for many people in many cultures, climate change and the needs of a rising human population take precedence over the needs of any non-human species.[7] Animals are food or for sport or for work. Animal welfare is seen as a middle-class luxury that can only be afforded by people who have plenty to eat and a rich lifestyle. It will be no good going to discussions on the future of the planet and making the case for animals based on sentiment or evidence that does not stand up to scrutiny. The place for animals will be hard fought and hard won. Claims about animals that turn out to be false or exaggerated will backfire in the end like 'crying wolf' once too often. Where the claims for non-humans apparently conflict with human ones (for example, a demand that animals should be kept in conditions that give less human food or make it more expensive), then the case will have to be particularly clear and watertight, because every excuse will be found to turn it down.

This book is structured around these two basic themes. One is animal welfare in the context of other world debates such as feeding people, human health, and the quality of our environment. The other is what we now know about the conscious experiences of animals and how this relates to their welfare. For reasons that will become clear, these are dealt with in reverse order so that the book ends with 'other' reasons (that is, reasons other than consciousness) for the importance of animal welfare and begins with a head-on confrontation with consciousness itself.

The book is aimed as much at people who think that the welfare of animals is not particularly important as at people who are already convinced that it is. To both groups, I want to show that a case for animal welfare that is based entirely on the assumption that animals have conscious experiences is a weak case, but it can be vastly strengthened with other kinds of evidence about the way animal well-being affects that of humans. Even though an anthropomorphic approach to animals may be appealing to those who are already convinced that animals have conscious experiences, it is limited in its capacity to persuade the large numbers of people in the world who are not convinced, or even those who are half-convinced but think that humans should always take precedence. It will not have sufficient weight against other competing claims, such as the need to ensure food security or to mitigate the effects of climate change. There are too many vested interests and too many cultural differences. Furthermore, the existence of conscious experiences in animals is increasingly being challenged by scientists offering 'killjoy' explanations of what had previously been taken as evidence for consciousness in animals. As if this were not enough, there is now new evidence from human neuroscience that shows just how difficult it is to understand consciousness at all, even in ourselves. So, the strongest arguments for animal welfare will come not come from pushing the case for animal consciousness beyond what the current evidence will support, but by linking the welfare of animals with that of humans. Self-interest is a powerful driver and it will be the most powerful ally that animal welfare can have.

Some people may find this conclusion disconcerting. Some may even object that, by looking at the evidence for animal consciousness with a critical eye, I am setting back the cause of animal welfare by several decades. On the contrary, I see making sure that arguments are sound and can stand up to scrutiny as a service to animals and animal welfare. Putting forward arguments that are most likely to convince those who are not already convinced is just what animal welfare needs most just now.

Perhaps I should say at this point that I am deeply and passionately pleased that the second half of the twentieth century saw such a spectacular change in attitudes to animals of all sorts. I am proud to have known people like Ruth Harrison,[8] who did so much to change public attitudes to farm animals, and Donald Griffin,[9] who brought a fresh light to the study of animal consciousness and forced many scientists to rethink their reluctance to study the subject. But I am also concerned that some of these welcome developments have also brought in their wake a degree of confusion, misunderstanding, and even antagonism to animal welfare. They seem to have given everyone a licence to talk about emotion, empathy, insight, and other human attributes in non-human species as if there were no limits, no rules any more; as if they were the same in other species as they are in us.[10] The trouble is that animals generate such deep and conflicting emotions in humans that it is often difficult to have the sane discussions that are needed to resolve these issues and so, in the end, to provide the best for the animals themselves.

This book is about freeing ourselves from these confusions by looking afresh at some key questions about non-human animals—how justified are we in projecting human emotions onto animals (Chapter 3), what kinds of mental lives (if any) do they have

(Chapters 4–6), what can science tell us (and what it cannot) about good welfare and quality of life (Chapters 7 and 8), and how can we improve the lot of non-humans in a world increasingly concerned about the human species and its future (Chapters 9 and 10)?

There are, of course, many different reasons why these questions generate such confusion and disagreement, such as culture, personal philosophy, and the rapid advances in science that not everyone knows about or believes in, but these do not seem to fully account for the wide range of views that exist about animals. Looking around for some other reason why people can hold such very different opinions about the same animals, time and time again the same culprits appear. Words. Language. The beautiful, ambiguous English language (and I am sure that other people can make a similar case for their mother tongues) gives us words to express poetry and an infinity of ideas. But many of those words also subtly direct our thoughts and lead us to make assumptions we may not realize we are making. Many words are loaded with so much baggage of their own that we fail to spot the effects they are having on us. So in rethinking attitudes to animals, and trying to see what questions we really need answers to, there can be no better place to start than with the words we use to describe them.

2

SEDUCED BY WORDS

I love words. I love the way they give us the ability to describe someone as 'mistaken' or 'lying', depending on whether we think they have simply got their facts wrong or have deliberately tried to deceive us. Just by using one word rather than another, we can completely alter what we are saying about that person's motives and personality, and make a friend or an enemy in an instant. But with the power of words comes danger. The word 'false' as an answer to the question: 'Is Paris the capital of Norway: true or false?' is a neutral description of something that does not match up to the real world. But 'false' in the context of a 'false friend', means someone who has let you down, made promises they did not keep, or actually betrayed you. Words are powerful but they are also fickle. They whisper their meanings and then change them again when you try to use them on another day. And it isn't just other people we influence by our choice of words. It is ourselves. Nowhere is this more apparent than when we use words to describe animals and what might be best for their welfare.

To an extent that you may not realize, our views about animal welfare are shaped by the descriptions we have heard and the words people have used. Take the word 'enrichment', for example. To enrich means to make rich, to enhance, to increase the amount of something that is valuable—all positive attributes. So if you hear that a zoo or farm is providing an 'enriched' environment for their animals, it is easy to assume—without any further thought—that they must be improving the welfare of their animals. Perhaps the zoo is providing food for the animals in a way that allows them to feed more naturally, or the farm is giving its pigs objects to play with. How could such enrichments not be good for the animals' welfare? Since they are 'enrichments', the animals must, by definition, be being given something valuable, something that enhances their lives. Who would prefer impoverishment to enrichment? Why should animals be different?

Of course, enrichments can be very good for welfare,[1] but just look at what happens if we change the emotional loading of the word we use to describe what is going on. Suppose that instead of calling it 'enrichment', we called it 'clutter', and instead of talking about 'natural feeding', we talked about 'preventing easy food access'. We would, with no difference in what was actually happening, have gone from something that clearly enhanced welfare to something that might not improve it at all and might even have a negative effect. At the very least, we need to look more closely at what the effect of giving something to an animal actually does. Does the animal use it or ignore it? Does it injure itself on the 'enriching'

9

branches? By attempting to eat the 'enriching' toys, does the animal poison itself or choke on the splinters? Simply by using different words, we open up a whole range of new questions and doubts that were simply not possible with the blanket assumption that all enrichments must be good for welfare. Once the word was used, it stopped the questions being asked.

You can play similar games with other words commonly used in connection with animal welfare. Try replacing 'free-range' with 'exposed to the elements', for example, and see how what looks initially like something that is overwhelmingly good by definition becomes much more open to question with a simple change of description. Conversely, 'intensive' sounds dreadful, whereas 'warm and comfortable and protected from weather' sounds a great deal better. Which is right? The point is that even if you end up concluding that free-range is better for animal welfare than intensive, the change of description at least draws your attention to the need to look into the evidence that it is better. Free-range means cold and wet and exposed to predators as well as some very real advantages.[2] Free-range systems have to stand up to scrutiny and show that they are better for welfare (if they are). They must not be allowed to hide their very real disadvantages by claiming to be the best by definition, and exempt from criticism just because the word 'free' in free-range resonates in such a powerful way with almost everyone. How can it not be good to be free? When basic needs for comfort and shelter are not met. Ask someone who has the freedom of the open road but no home and no shelter.

The advertising industry is, of course, well aware of this power of words to bypass our rational thinking and persuade us that a product is good because it is 'natural', or comes from 'a country kitchen', or has some other attribute that resonates in a hidden corner of our minds. City dwellers in particular are susceptible to a romantic notion of what they think an idealized farm is like, where a small group of chickens scratches around a farmyard and a farmer's wife collects the eggs, still warm, from under each hen by hand. That is where they would like eggs to come from and if there is just a hint in the way eggs are described on the box that they really were produced in this way, then that becomes an immediate selling point. People buy, and they feel good about buying, because they are helping that farm of their imaginations.

The word 'natural' has a particular power to numb thought and, like some love potion, to lead people to see everything it touches as unquestionably good. 'Natural fibre', 'natural essence', 'naturally grown'—these need no further accolade. Their desirability and even their safety seem to be guaranteed by their association with this one word. 'Natural' is so often associated with 'goodness' that other words (and products) take on the reputation of goodness, just by association. Words resonate in the mind because of the company they keep.

Interestingly, these words of power can be quite specific and distinct from even closely related words. So while the word 'natural' conjures up some (probably fictitious) rural idyll before 'civilization' took over, the word 'nature' has taken on a quite different meaning and become in many people's minds associated with Tennyson's 'nature red in tooth and claw'. 'Nature' is fierce and bloody and not at all benign. Nature has predators and starvation and disease, not at all like the gentle 'natural' world—until you stop to think that they are both describing the same thing. Words do literally seduce us. They sweep us up in their romantic aura and stop us seeing reality, or at least they give reality a very special 'spin'. For that, of course, is what politicians and campaigning organizations do all the time. They choose the words that not only bypass thought but even make us believe we have already done the necessary thinking and have arrived, by a completely rational pathway, at our well-reasoned conclusions. In fact, on so many occasions, the words we have been presented with have led us to cut corners, gloss over the inconvenient facts, and unthinkingly assume that something is good, desirable, or even morally justified. That brings us to the biggest danger of words. They don't just give us imaginative mind-pictures. They can also influence what we think of as right or wrong, ethically good or morally reprehensible. They can turn something into the most desirable ethical goal or condemn it as the worst form of depraved behaviour. Just by which one of them we choose to use.

Sometimes the moral baggage that a word carries with it is obvious, which makes it powerful but not particularly dangerous. The word 'cruelty', as part of its definition, implies something morally bad, something we should not do. So if we say that a person is being cruel, we use that word to condemn their action and to convey our view that they should be stopped and punished. But if we discovered that what we thought was a cruel action (stabbing an animal with a knife, say) turned out to be a life-saving surgical operation, we might stop using the word cruel altogether. The word 'cruel' is thus clear enough in its meaning that it can be picked up or put down again as we see fit. It wears its moral credentials on its sleeve so that we are aware of what we are doing when we use it. The same action, in this case cutting an animal with a knife, can be cruel or not cruel depending on circumstances. The word can be clearly separated from the action it is applied to.

The same is not true of other words, such as 'vivisection' or even, in some peoples' minds, 'experiments' when applied to animals. Such words are so loaded emotionally that it is almost impossible to separate what they are describing from a moral view of the rightness or wrongness of what they are describing. Strictly speaking, vivisection just means 'cutting living things', so an operation performed by a veterinary surgeon for the benefit of an animal would, on this definition, be vivisection—a living animal is cut. But most people would not want to call veterinary operations 'vivisection' and they would not want to call vets 'vivisectors' because 'vivisection' has come to have such negative moral overtones.

Vivisectors are, by some peoples' definition, cruel. By definition, they perform operations without anaesthetics and take pleasure in inflicting cruelty on animals. By definition, they are morally reprehensible. The word 'vivisection' has shifted from its original meaning of 'cutting living things' to now carrying the implication that animals suffer horribly because they are cut open without anaesthetic or proper post-operative care. An operation carried out for research purposes, rather than veterinary purposes, could be described as 'vivisection' and condemned as immoral. But think about it for a moment. A veterinary operation, performed to save the life of an animal, might involve exactly the same operation, with exactly the same anaesthetics and the same care for the animal as a research operation, yet only the latter would be described as 'vivisection' and only the latter would carry the moral outrage.

Now there are certainly important differences between carrying out an operation to save the life of an animal and carrying out a very similar operation for research purposes, even when the research might directly benefit humans or other animals. The 'who benefits?' argument could put the same procedures in very different moral positions. But in deciding whether or not a given operation is morally justified, it is important not to confuse the argument about who benefits, from the equally important arguments about what actually happens to the animals. A given operation could be carried out with or without anaesthetics, with or without surgical skill, with or without post-operative care. There could even be a veterinary operation carried out without anaesthetic or post-operative care, and 'vivisection' carried out with anaesthetic and full care for the animal throughout. You might want to object to both or to either, but the grounds for objecting would be quite different. The operation benefits the animal in one case but causes it to suffer. The operation minimizes suffering in the other case but does not benefit the animal. Condemning only the second because it carries the label 'vivisection' would obscure the real welfare issue and, indeed, get in the way of reducing the animal suffering that might well be occurring in the badly supervised veterinary operation.

As it is, the widespread use of the term 'vivisection', with its massive negative overtones, has the effect of turning many people against science and against research, not because they know and condemn what actually happens but because they are, quite naturally, against something as terrible as 'vivisection' implies. This has the unfortunate effect that someone who takes a tiny drop of blood from an animal (no more than the pinprick a doctor might take from your finger for a blood sample) becomes a 'vivisector', a cutter of living things, and therefore to be stopped from inflicting cruelty on animals. He isn't given a chance to explain how small the pinprick was, or how the animal hardly responded. By calling him a vivisector, when all he has done is to take a small drop of blood, attention is drawn away from the serious cases where unnecessary suffering might really be being caused, because everyone is tarred with the same brush. By labelling everyone who works on animals as vivisector, the real issues are obscured. The animals don't benefit because the serious debates we need to have about the welfare of animals used in research are instantly and dangerously polarized by hurling the accusation of 'vivisector' and relying on the terrible reputation of the word to do its damage on anyone who would rather be carried along on an emotional wave than think for themselves about what it actually means.

The word 'vivisection' has such a powerful aura of authority that it is even able to spread its influence to otherwise neutral words such as 'experiment'. 'Experiment' on its own means nothing more than 'test' and is usually used in the context of trying out or testing a hypothesis. It literally means to know something 'from experience'. On its own it says nothing at all about what sort of experiment or what sort of test is involved. But couple it with animals as 'experiments on animals' and it becomes a ready-made recipe for some people to raise very serious moral objections just to an idea, without seeing the need for any further information at all. A perfectly good word has been hijacked and its original meaning distorted. The experiments in question might involve the experimental testing of a drug, but, equally, they might consist of no more than experimentally offering cats a choice of different sorts of cat food to see which one they prefer. Any statement to the effect that eight out of ten cats prefer one sort of cat food to another implies that an experiment has been done.

An experiment aimed at finding out what cats really like to eat is of course very different in all sorts of ways from an experiment that involves surgery, but it is an 'experiment' nevertheless. So to condemn all 'experiments on animals' equally, just because they are, technically, experiments, is simply bad logic. It does not help animals to lump all experiments together as all equally likely to cause suffering, all equally likely to involve 'vivisection', and all equally legitimate targets for protest because it distracts attention from what really might need to be changed. It alienates people who are making genuine efforts to reduce any suffering or even discomfort to animals because they will still be labelled as 'experimenters' or even 'vivisectors', despite not cutting anything at all. The phrase 'experiments on animals' is so emotionally loaded that, for some people, the argument is over before it has even begun: bad cannot be good.

The mistake here is to believe that the moral judgement of what is right or wrong is part of the definition of the word 'experiment' itself rather than to separate the facts about what actually happens from the decision to approve or disapprove. But if we are to think clearly about animals and what is best for them and for us, we must make this distinction. We must separate what is from what ought to be. We must not be seduced into blurring this distinction by the powerful overtones that words carry or by the gloss put onto them by the media or campaigners or anyone else with strong views of their own who want to change the way we think. One man's freedom fighter becomes another man's terrorist. But words should be the servants of our thoughts and opinions, not the masters. We need to be eternally vigilant to the subtle effects they can have, often without our realizing it, first on our emotions and then, when their grip here is secure, on our most deeply held convictions of what is right and wrong. We need to be particularly aware of this power of words to subvert, direct, and influence our judgements when it comes to our dealings with non-human animals because they arouse such strong emotions in so many of us, and we are therefore particularly susceptible to the implications they carry.

But if words like 'free' and 'experiment' can influence us emotionally, either positively or negatively, at least we can identify them as different words describing different things and likely to have different effects on us. We can, if we take the trouble, see what emotional loads they are carrying around with them and we can, if we are strong-minded enough, try not to be too influenced by them.

There are, however, some words that have even more important effects on the way we think and are even more subtle in the way they do it. The problems these words cause are not so easily dealt with by substituting one word for other, less pejorative ones, because often we don't have alternative words to use. The result is that different people use exactly the same words but mean totally different things by them. They 'talk past' each other,[3] assuming that because they are using the same words, they must be talking about the same thing, and not realizing the depths of the differences between them. This is only partly the fault of the words themselves, slippery and ambiguous though they may be. It is much more because we use words that we barely understand the meaning of ourselves—words like 'mind', 'emotion', and 'consciousness'. It is bad enough saying what we mean when we apply such words to humans. So applying them to animals—or even asking whether they could or should be applied to animals— leaves us hanging by our fingertips in an even more desperate attempt to understand what this might mean.

17

We don't have any words to express what it would be like to have minds that are not like human minds, or emotions that are not like human emotions, or experiences that are conscious but not like human conscious experiences.[4] All we have are the words that we use to describe our own human experiences. So we use them and, by using them, deliberately or not, convey the meaning that the human experiences the words describe are the same as those in animals. Some people are quite happy with this implication. It is what they mean to say.[5] They believe it is not even worth questioning whether animals have conscious experiences like us because they so obviously do. They also believe that the best way to understand animals is through our own conscious experiences. This is called 'anthropomorphism'—seeing animals in human form. Others, however, are more wary. They do not wish to be pushed by an inadequacy of words into an inadequacy of ideas. They want to try to understand animals for themselves and believe that anthropomorphism does nothing but blur important distinctions between humans and non-humans.

The scene is now set. The battle lines have been drawn. In various ways, we can see that words are often a hindrance rather than a help because they convey such different meanings to different people. To see why animals matter, we have to go beyond the emotional pull of these powerful words and look rationally at some basic ideas about animals themselves.

3

THE TROUBLE WITH ANTHROPOMORPHISM

'Anthropomorphism' is the name given to the almost universal human tendency to attribute human qualities to things that are not human. 'The cruel sea' or 'the selfish gene' are examples of anthropomorphism used for dramatic effect or as a shorthand way of expressing a complex idea. Some religions do, of course, see the sun, the wind, and other inanimate objects as having minds of their own, but most people would say that it is 'as if' they had human qualities, not that they actually do.

When it comes to animals, anthropomorphism is often used quite specifically to mean attributing human mental experiences to animals[1] and it is here that the 'as if' part of anthropomorphism becomes more controversial. Do animals behave 'as if' they had thoughts and feelings like us or do they really have them? Are dogs 'happy' or really happy, as Marc Bekoff insists we should say?[2]

The key issue is whether anthropomorphism simply provides us with some useful metaphors or whether it actually tells us about animal minds.

There is no doubt that anthropomorphism can often be extremely useful in science. In physics and chemistry, the idea of objects or molecules 'wanting' to do things can be a great help in visualizing how they might behave or where they might come to rest. But it is in the study of animal behaviour, and particularly the behaviour of primates, that anthropomorphism has been put forward as an actual method for obtaining scientific data. 'Anthropomorphising *works*', say Dorothy Cheney and Robert Seyfarth in their insightful (and scientific) study of vervet monkeys.[3] 'Attributing motives and strategies to animals is often the best way for an observer to predict what an individual is likely to do next.' Many others who study the behaviour of primates in the wild or in natural groups emphasize the enormous value of attributing human-like qualities to the animals they are watching. Franz de Waal, for example, was drawing on his long experience of studying the behaviour of chimpanzees when he argued that if we refuse to see the similarity between human and animal emotions we risk missing something fundamental, about both animals and us.[4] And Robin Dunbar[5] echoed the same idea when he insisted that 'we need to determine not just what an animal does, but, more importantly, what it is *trying* to do'.

In his pithy little book *The New Anthropomorphism*,[6] the late John Kennedy made the important distinction between using anthropomorphism as a crutch, as a useful way of thinking about a problem (what he called 'mock anthropomorphism'), and genuine anthropomorphism—that is, assuming that animals have mental experiences like humans. It is in the mock, metaphorical sense of anthropomorphism that we talk about water 'wanting' to find its own level, or molecules wanting to join together, or genes being selfish, and it is in this same sense that Cheney and Seyfarth make it clear that they find anthropomorphism to be useful, if not essential, in understanding primate behaviour. But others, most notably Marc Bekoff, go in for full-blooded, genuine anthropomorphism. 'Animals don't merely act "as if" they have feelings. They have them', he writes.[7] And even more forcefully: 'My colleagues and I no longer have to put tentative quotes around such words as *happy* or *sad* when we write about an animal's inner life. If our dog, Fido, is observed to be angry or frightened, we can say so with the same certainty with which we discuss human emotions.'[8]

The problem with this very overt kind of anthropomorphism is not that it is wrong (it may well be right) but that it leads to a complete lack of scientific rigour in the way we look at animals. It opens the floodgates to pure and unbridled speculation, with no constraints and no ground rules. Take a story that Bekoff cites as an example of how anthropomorphism can be used to study the feelings and thought processes of an animal.[9] A woman called Norma Harris from Texas wrote to him saying that she and her husband had noticed a squirrel coming out of a hole in their attic. Not wanting to be infested with squirrels, they had closed the opening. Later that morning Mrs Harris was sitting on the other side of the house when a squirrel peered in at the window and started chattering. It then slowly stood up and raised its front paws. Mrs Harris continues: 'She was showing me two rows of prominent black nipples and her full breasts! Then I realized we had locked her away from her babies in the attic.' Mrs Harris was convinced that the squirrel was asking her to unblock the hole. Using this story as an example of how to study animal behaviour is virtually tantamount to saying that there are no limits to how we interpret animal behaviour. Anyone can have a go and attribute anything they feel like to animals—such as squirrels with an ability to reason and a belief in the altruistic tendencies of human beings. And why stop at squirrels that deliberately use their lactating state to persuade human beings to do things for them? If there are no limits, the possibilities are, literally, endless.

It is important to realize that anthropomorphism was, for much of the twentieth century, totally rejected as part of the study of animal behaviour for precisely this reason—that it leads to the unwarranted and untestable attribution of human characteristics to other animals. Over the course of that century, the study of animal behaviour grew in the capable hands of Niko Tinbergen, Konrad Lorenz, and other pioneers, from its beginnings in natural history into a full-blown science called ethology. And one of the things that helped ethology to become a respectable science was its rejection of anthropomorphism, the fact that it set out to be as objective as possible and to describe the behaviour of animals without implying that they were little furry or feathery human beings. Of course, the ideal of being objective was often hard for ethologists to live up to precisely because everyone finds it so tempting to slip into the language of animals 'being angry' rather than simply describing what they are doing, but gradually the idea that this was what should be aimed at became widely accepted. Anthropomorphism was to be avoided, not because animals have no feelings but because trying to study those feelings was fruitless. 'Because subjective phenomena cannot be observed objectively in animals,' Tinbergen wrote in *The Study of Instinct* in 1951, 'it is idle either to claim or deny their existence.'[10]

This is an important point. Tinbergen was quite open to the idea that animals might have feelings, possibly even feelings like human ones. He says quite clearly and explicitly later in the same book: 'I do not want to belittle the importance of a study of either the directiveness of behaviour or of the subjective phenomena accompanying our and possibly the animals' behaviour.'[11] But he then goes on to explain the problem with trying to study them: 'Hunger, like anger, fear, and so forth, is a phenomenon that can be known only by introspection. When applied to another subject, especially one belonging to another species, it is merely a guess about the possible nature of the animal's subjective state.' It was this guesswork nature of anthropomorphism, not the existence of subjective feelings, that Tinbergen was objecting to. And it was anthropomorphism as a method, not the belief that animals lacked conscious thoughts and feelings, that led the early ethologists to reject it.

This battle against anthropomorphism was a long and hard one and it raged through much of the twentieth century. The main opponents were called 'behaviourists', because they insisted that everyone should be studying observable behaviour rather than unobservable feelings. Behaviourists adopted a variety of different positions, some of them much more strident than Tinbergen's mild and rather polite way of acknowledging, but then refusing to study, animal feelings. J. B. Watson, for example, one of the chief proponents of behaviorism, used extremely colourful language to ridicule the whole idea of talking about consciousness in either animals or people. 'States of consciousness, like the so-called phenomena of spiritualism, are not objectively verifiable,' he wrote in no uncertain terms, 'and for that reason can never become data for science.'[12] To Watson, studying consciousness was not just idle guesswork. It was simply not respectable at all, on a par with a belief in fairies. 'One can assume the presence or absence of consciousness anywhere in the phylogenetic scale without affecting the problems of behaviour by one jot or tittle and without influencing in any way the mode of experimental attack upon them,' he explained as the essence of his objection to consciousness.[13] Consciousness was untestable. There was no way in which an experiment could be devised that would have one outcome if the animal were conscious and a different outcome if it were simply behaving 'as if' it were conscious. Not 'one jot or tittle' of a difference. And if there was no difference, even in principle, then it was not science.

Watson and other influential behaviourists such as B. F. Skinner continued to mount such sustained attacks on consciousness that behaviorism became, in many people's eyes,[14] not just a cautious approach to a difficult subject, but a categorical denial that the study of animal feelings had any meaning at all. Even John Kennedy, one of the most outspoken critics of all forms of anthropomorphism in biology,[15] describes Watson and Skinner as having gone 'too far' in their rejection of internal causes of behaviour, including subjective feelings. 'Not scientifically respectable' had become 'does not exist'.

That was why the publication of Donald Griffin's book *The Question of Animal Awareness* in 1976 had such an electrifying impact on the study of behaviour.[16] In striking contrast to the prevailing behaviourist views of the time, Griffin argued passionately that animal awareness was, or at least should be, a suitable subject for scientific study. He urged ethologists to throw off what he called the behaviourist 'taboo' and start trying to think of ways to make animal consciousness scientifically respectable. He was aware of how difficult this was going to be, but thought that animal communication, particularly primate communication, might be a 'window' into animal minds.

Part of Griffin's impact came from the fact that he was a distinguished biologist who had not previously been known to harbour anthropomorphic tendencies. His reputation had been built at the 'hard' end of biology—the physics of bat sonar and bird migration. Here was no soft or woolly minded biologist being anthropomorphic because he couldn't cope with hard science, but a highly respected scientist calling for a new way of looking at animals. The conscious experiences of animals, for too long banished altogether, needed, in his view, to be brought back into mainstream biology, even if we had to be a bit anthropomorphic to do so.

Griffin changed the way people thought about animal consciousness, although possibly not altogether in the way he intended. Following the publication of his book, a whole new branch of ethology, cognitive ethology, was born. For the first time for many years (since the beginning of the twentieth century, in fact, when the behaviourist attacks really got going), some degree of anthropomorphism was to be allowed as a method of studying animal behaviour. With care, and with caution, it began to seem possible to bring the study of animal consciousness into biology and make it a respectable part of how we study animal behaviour.[17] The trouble was that no one specified the conditions on which it was to be allowed back. There were no ground rules for when anthropomorphism could be used legitimately and when it had to give way to more accepted scientific methods. So once it was

conceded that anthropomorphism might have some role in understanding other species, this was taken by some people as meaning
that all behaviourist caution could now be thrown to the winds.
The taboo was broken and, as far as discussing animal feelings was
concerned, there were now no limits.

A few lone voices, such as Gordon Burghardt, tried to argue that
the use of anthropomorphism in ethology should be restricted to,
say, generating new hypotheses that would lead to testable predictions about behaviour.[18] But the floodgates had been opened and
any such boundaries were soon swamped. 'To live with a dog is to
know first hand that animals have feelings. It's a no-brainer,' wrote
Marc Bekoff.[19] It began to look as though no further thought or
investigation were going to be necessary. Even worse, this new wave
of anthropomorphism threatened the very scientific basis of the
study of animal behaviour itself, particularly that branch of it known
as cognitive ethology. Anecdote, analogy, and anthropomorphism
were put forward as the basis for reaching conclusions.[20]

Anthropomorphism had become not just an extra dimension to
the study of animal behaviour, but one of its central methods. It
was as if a guest invited in from the cold to a banquet had then
taken over the castle and proclaimed himself king. As Wynne put
it,[21] by agreeing to take back the baby (consciousness), ethology
seemed to be forced to take back all the dirty bathwater as well.
Some behaviourists may have gone too far in their rejection of
animal feelings but does this mean we now have to choose between
rejecting feelings altogether and allowing anthropomorphism to
run riot, aided and abetted by some startling anecdotes and a few
colourful analogies? Are there to be no standards, no objective
ways of testing hypotheses? Are ethologists now to be ridiculed for

daring to question anthropomorphic interpretations of animal behaviour on the grounds that it shows they don't care for animals or that they are denying the possibility of conscious experiences in other species?

Just as behaviorism went too far in its day and the tide then turned against it, so now anthropomorphism has gone too far the other way, as Kennedy saw with great clarity.[22] If left unchecked, it is in danger of rotting away the very fabric of ethology. The central lesson of behaviorism—that subjective experiences are private and cannot be observed directly with the same methods that we can objectively observe behaviour—remains true and we forget that at our peril. Whatever 'taboos' are broken, it remains stubbornly true that two people can observe behaviour and agree on what they see, but those same two people cannot observe the private subjective experiences of one of them and agree on what is being experienced because only one of them has access to what is going on from the inside. However loudly we declare that consciousness is part of biology, the study of consciousness cannot use the same objective scientific methods of hypothesis testing that are used in the rest of biology. This point will emerge more clearly in Chapter 4.

Jane Goodall, someone who has used anthropomorphism very successfully in her own work, is very clear about the distinction between being scientific and being anthropomorphic. She describes it as like looking through different windows. For her, there are some windows that are 'opened up by science, their panes polished by a series of brilliant, penetrating minds'. She continues: 'But there are other windows, windows that have been uncluttered by the logic of philosophers, windows through which the mystics seek their visions of the truth.'[23]

The idea of different windows on the truth may be the easiest way of reconciling the desire of ethologists to be objective and to work with testable hypotheses, and the insistence of other people (often including those same ethologists and philosophers) that it is just plain obvious that animals have conscious experiences even though we cannot test this scientifically. You can be anthropomorphic and look through one window, being completely certain that your dog is experiencing pain or pleasure although not being able to prove it scientifically. Or you can be scientific and look through another window, demanding evidence that can be tested, questioning current orthodoxy, thinking of alternative explanations, and looking for all the world like a sceptic about animal feelings. There is nothing contradictory about seeing one thing through one window and something else through another, but it is important to be aware that you are getting different views of the world.

So why exactly is anthropomorphism not good science? What is wrong with the telling anecdote or the intuitive leap of analogy? Let us start with anecdotes, one of the mainstays of anthropomorphism. Anecdotes can be an extremely useful starting point for any scientific investigation, an indication of what needs explaining or what cannot be explained by current theories. A good example of the constructive use of anecdotes in ethology is a paper by Dick Byrne and Andy Whiten in which they collect together many different anecdotes about monkeys and apes apparently deceiving each other.[24] One of the anecdotes they describe is a famous report by Hans Kummer from his many hours spent watching the behaviour of baboons in the wild. One day, Kummer watched as a female baboon gradually edged her way so that she was sitting behind a rock and almost hidden from the dominant male in the group, who could only see her head. Once behind the rock, she proceeded to groom a subordinate male, her hands out of sight of the dominant male who would have attacked her had he seen her doing this. The fact that the female took about twenty minutes to reach the safety of the rock, moving her bottom an inch at a time, suggested to Kummer that she was deliberately deceiving the dominant male by not letting him know what she was doing and making sure she could groom the other male out of his line of sight behind the rock.

Kummer's anecdote shows how complex baboon behaviour can be and raises the possibility that baboons might indeed be capable of deliberately deceiving each other. One story on its own, however, does not constitute evidence that baboons can really do this, not because it is a single incident, but because it does not tell us enough about the female's behaviour. She might, for example, have previously learnt by trial and error over a long period of time, that behind that particular rock was a place where she did not get attacked by the dominant male, without understanding why. Our human interpretation that she worked out what the male could or could not see and deliberately moved out of his line of sight is not the only plausible explanation of her behaviour. This means that even when this story is put together with all the other anecdotes that Byrne and Whiten were able to collect, it does not constitute evidence for deliberate deception any more than seeing one hundred conjurors performing the same card trick constitutes more evidence for 'real' magic than seeing just one performing it. All the conjurors could be using the same entirely non-magical sleight of hand. What is wrong with anecdotes is not that they are one-offs (one well-authenticated sighting of a living dodo would be enough to say that the species was not, after all, totally extinct[25]). What is wrong with anecdotes is that they are not experimental and they leave us with several different hypotheses of how the behaviour is really produced. Anthropomorphism tends to make people go for the most human-like explanation and ignore the other, less exciting ones.

The example of Clever Hans—the counting horse who turned out not to be counting after all—has been cited many times as an example of how misleading anthropomorphic interpretations of 'clever' animal behaviour can be.[26] But I am going to use it again here because the experiments carried out by Otto Pfungst to show that the horse was not in fact counting were so ingenious and so cleverly devised that they still serve as a warning about the dangers of anthropomorphism for us today. The point of the story is not to show that horses can't count. It is hardly surprising if a horse does not turn out to be a mathematical genius. The point of this example is to show how extraordinarily difficult it was to demonstrate that he was not counting at all, and even more difficult to work out what he was really doing.

Clever Hans was a horse who lived in the early part of the twentieth century. His owner, Wilhelm von Osten, made a considerable living by taking his horse to fairgrounds and getting people to ask the horse questions, such as what is the sum of two plus two. Hans would give the answer by striking his foot on the ground the correct number of times and then stopping when he got to the right answer. The horse's fame spread. Eventually, stories of a horse that could count, do addition and subtraction, deal with fractions, remember the names of different people, and read German reached across Europe and the United States. A commission of inquiry into Clever Hans's abilities concluded that no tricks were involved and so his owner was not deliberately deceiving anyone. However, at least one person, Oskar Pfungst, was unconvinced by all the stories of what the horse could do and began to investigate what was really happening. Clever Hans was certainly impressive. It was possible to ask him a wide variety of questions which, provided they could be answered by a hoof striking the ground, Hans was consistently able to answer correctly. He could be asked questions verbally or he could be shown a blackboard with the question written on it in chalk. It didn't seem to affect the accuracy of his answers at all.

What did matter was whether he was in sight of a person who knew the right answer. This didn't have to be his owner, which ruled out fraud, but if he was isolated from a knowledgeable person or was made to wear blinkers so that he could not see them, his powers failed him and he was unable to give the right answer. The most critical part of what Pfungst did was to vary whether the person standing next to Hans did or did not know the right answer themselves. He devised an experiment (one we can still learn from today) in which sometimes von Osten did not know which question Hans had been asked or was deliberately told the wrong answer. For example, horse and owner might be shown different blackboards each with a different question. Pfungst found that Hans consistently gave the answer his owner thought was the correct one rather than the answer that was actually correct. If von Osten thought the question was 2 + 3 and Clever Hans had been asked for the sum of 2 + 1, Hans would hit the ground five times, not three. This clearly showed he was somehow taking his cue from the owner, but it was not clear what the cue could be. Herr von Osten was unaware that he was doing anything at all and was genuinely under the impression that his horse was a mathematical genius.

After numerous other experiments, Pfungst was able to show what Clever Hans was really doing. It turned out that when the horse was asked a question and began to strike the ground, von Osten inadvertently tensed himself slightly as the horse got near the right answer and then relaxed with a slight release of breath when he had done the right number of hoof strikes. This tiny change in his owner's behaviour, almost imperceptible to other humans, was enough to tell Hans he had to stop hitting the ground. Hans was so good at this that he could pick up cues not just from his owner but from other people as well, even those unfamiliar to him. Hans was certainly clever but not in quite the way that had been claimed.

The interesting thing is that the real explanation of Hans's behaviour is complicated and not simple at all. He had to notice the connection between hitting the ground with his hoof, an almost imperceptible change in human behaviour, stopping hitting the ground with his hoof, and being given a lump of sugar or a piece of carrot. Counting seems easy by comparison, so the simplest explanation isn't necessarily the right one. We owe Otto Pfungst a debt of gratitude for not being satisfied with anecdotes about a counting horse and for insisting on conducting experiments to find out what was really going on. His elegantly controlled trials taught us more about animal behaviour than a hundred anecdotes of counting dogs, pigs, or horses and still stands as an exemplar of how to conduct experiments on 'clever' animals.

So, anecdotes have a valid place in ethology, not as evidence in their own right, but as starting points for hypotheses to be tested later by experiment. Anthropomorphic interpretations may be the first ones to spring to mind and they may, for all we know, be correct. But there are usually other explanations, often many of them, and the real problem with anthropomorphism is that it discourages, or even disparages, a more rigorous exploration of these other explanations. Rampant anthropomorphism threatens the very basis of ethology by substituting anecdotes, loose analogies, and an 'I just know what the animal is thinking so don't bother me with science' attitude to animal behaviour.

But scepticism, if properly applied, can also make hypotheses about animal consciousness more, rather than less, likely. It sounds paradoxical, but doubting whether animals have thoughts and feelings, and then systematically eliminating alternative, non-conscious explanations of their behaviour, could in the end make the conscious ones more plausible.

My friend the late David Wood-Gush used to fantasize about devising an experiment in which, if a pig were conscious, it would turn left, and if it were not, it would turn right. He never succeeded. He, like everyone else, was always left with the frustrating fact that he could think of no clear difference between what a conscious pig would do and what a pig just behaving 'as if' it were conscious would do. Both would turn left or right at exactly the same time, and not 'one jot or tittle', as Watson would say, would distinguish their behaviour. But my goodness he did try. He didn't just assume that his pigs would solve whatever problem he had just come up with in the same way he, as a human, would. His fantasy experiments had controls for everything—taking cues from humans, simple learning techniques, smell, etc., so that more and more hypotheses could be eliminated. He was a sceptic par excellence but he used his scepticism to try to demonstrate the consciousness of pigs by thinking of an experiment that would eliminate all the alternatives he could think of. He believed that if he could do that, he would make the hypothesis that pigs were conscious the most plausible one around. The catch, which of course he recognized, was that it was impossible to eliminate all the other hypotheses with complete certainty. Someone could always say the pig was behaving 'as if' it were conscious but it wasn't really. Behaviorism would always strike again.

What was impressive about David's unending quest for the perfect experiment was not his failure to show how to demonstrate conscious experiences in pigs, but his determination to go as far as he could in eliminating as many of the alternative explanations as possible. Yes, he knew that at some point he would have to drop his scepticism, leave the strictly scientific behind, and make an anthropomorphic leap into pig consciousness. But he planned to make the smallest leap possible. He didn't just assume that pigs are just like humans and jump from human consciousness to pig consciousness in one giant step, leaving himself open to countless charges of ignoring other equally plausible hypotheses that did not involve consciousness. The whole aim of the fantasy experiment was to systematically undermine these other explanations, leaving conscious pigs as more and more plausible as each alternative bit the dust.

Scepticism, hypothesis testing, even the search for simpler explanations are not, therefore, the hallmarks of people who want to deny the existence of animal consciousness altogether, but in many cases they are the precise opposite. They are essential steps towards showing that conscious thoughts and feelings are the most likely explanation of behaviour. They are the lifeblood of science and are the reason we put our trust in some of the most basic things in life. Every time you drive a car that has been safety tested, drink a glass of water that has been analysed for quality, or visit a doctor who has passed examinations, you are benefiting from someone else's scepticism. Someone doubted that the car, the water, or the doctor were up to scratch and tested them. The fact that they passed the sceptic's tests is what gives you confidence that they are trustworthy.

So it is with claims about animal thoughts and feelings. We are more likely to believe that an animal is genuinely working something out in its head or really counting if the simpler explanations have already been ruled out. If it had turned out that Clever Hans hadn't been taking cues from his owner and no one could explain his behaviour except by assuming he could do mental arithmetic, we would all have been really impressed. And think how delighted those with an anthropomorphic way of approaching animals would have been. The sceptics have been proved wrong, they would have said. Precisely. A good sceptical approach has been used to rule out alternatives. Just as it should be in science.

The critical, doubting, sceptical methods of science are thus the friend, not the enemy, of the idea that other animals may have thoughts and feelings a bit like ours. Anthropomorphism, by discouraging and ridiculing such scepticism, does a disservice not only to ethology but to a wider acceptance of consciousness in other species. Throwing off the chains of behaviorism (breaking the taboo) may not be as fruitful and liberating as it seems at first if the result is a few shaky anecdotes for which there are plenty of non-conscious explanations that haven't even been tested. Cognitive ethology could easily become a laughing stock to the scientific community if it allowed anthropomorphic thinking to take over and came to regard scepticism as a dirty word or an indication that someone was denying animal consciousness altogether. We need all the scepticism we can muster, precisely because we are all so susceptible to the temptation to anthropomorphize. If we don't resist this temptation, we risk ending up being seriously wrong.[27]

There is already a whole industry devoted to trying to make us seriously wrong. Its aim is to exploit our anthropomorphic tendencies and make us respond to machines 'as if' they had feelings and emotions, just so that we will be more likely to buy them. Computers and cars are being built to trick us into thinking we could have a relationship with them. For example, a recent paper with the intriguing title 'Is that car smiling at me? Schema congruity as a basis for evaluating anthropomorphized products'[28] shows that commercial interests are on to us. They have rumbled our tendency to anthropomorphize everything and are systematically exploiting it. Even people who work with computers and design them seem to be unable to resist the temptation to attribute human feelings to their machines.[29]

This anthropomorphizing of machines has major implications for our understanding of animals and should make us even more sceptical of the anthropomorphic approach to them. The most extreme form is to be found in the computer pets that are specifically designed to get people to interact with them as if they were real animals. This began in the 1990s with the development of toy pets operated by a computer, such as the Japanese Tamogotchi. The Tamagotchi interacted with its owner in various ways and could express 'emotions' by switching on lights of different colours. The owner had to respond and care for it in certain ways or it would 'die'. Modern technology, however, can do even better by making digital pets that behave more and more like real animals. Take this description of a digital pet tiger cub:

> She was born in the wild, and now she needs your love and care! She's very sleepy, so wake her up gently! Give your newborn lots of love and affection by petting the soft fur on her back. Then watch your baby cub sit up, open her eyes and make cute baby cub sounds … or even give a little roar! Keep petting, and your cub becomes more playful. But if you stop petting her, she'll lie her head down and go back to sleep. The more you love your cub, the more she'll respond to you! Cub figure comes with a bottle and special adoption certificate. Includes 3 'AAA' batteries.[30]

If it weren't for the giveaway point about batteries, this could be a description of how to take care of a real animal.

The advanced video technology developed to make realistic games has also been turned to producing digital pets that are specifically designed to make people think they are interacting with a real live animal. Owners can only understand their pet if they watch its behaviour, body language, and facial expressions, just as with a real pet. They can play with their pets and train them to do different things. The pets will respond to 'touch' and their behaviour is deliberately made as natural as possible and slightly unpredictable so that the owners develop a sense of having a relationship with them. To increase this sense of a real relationship even further, the owner's behaviour is often made to have a long-term effect on the pet's health and well-being; for example, by playing with it regularly, the pet becomes more interactive, 'optimistic', and likely to live longer.

This is of course just the beginning. As technology advances, there will be robo-pets that look, sound, and feel much more like real animals, and that behave like real animals too. We will stroke them and interact with them and they will respond 'as if' they were pleased to see us. Bekoff argues that the relationship we build up with real dogs, cats, and other animals is the basis of the anthropomorphic claim that these animals have feelings like us.[31] His argument is that people would not feel such a deep emotional attachment unless the animals themselves really felt emotions. This is a powerful argument and, as a dog-lover myself, I have felt its force. But robo-pets that mimic many or all of the behaviour of real animals reveal the weakness of this argument. Anthropomorphism alone could lead us to the conclusion that anthropomorphized products need to be treated like real animals or real people.

In a world where it is already hard to persuade people to give enough priority to the welfare of real animals, do we really want to be sidetracked by unscientific approaches to animals that leave us with no more understanding of them than from our own intuitions? The sceptic, the person who questions whether behaviour really reveals feelings, or only seems 'as if' it does, looks a lot less foolish and a lot more constructive when confronted with a robo-pet giving a near-perfect imitation of a real animal. It may be 'bad biology to argue against the existence of animal emotions',[32] but is it bad computer science to argue against the existence of robo-pet emotions? And if it is, what is the difference? Don't we owe it to real animals to be a bit clearer than the vague anthropomorphic feeling we get from playing with our dog (especially if it's a robo-dog)?

One of the reasons anthropomorphism is unprepared for the ethical dilemma of how to treat a digital pet that behaves like a real animal is that it underestimates the power of simple mechanisms to produce complex behaviour and to 'fool' us into thinking we are looking at a conscious organism. Or, to put it another way, it sees man-made machines as quite incapable of producing behaviour remotely as complex and versatile as that of real animals so that there is no risk of confusing the two. It is a widespread belief that animals must have conscious experiences because their behaviour is so adaptive and flexible. Jonathan Balcombe argues that they cannot possibly be unconscious machines because 'machines are not flexible'.[33] Marc Bekoff makes the same point: 'Animals display flexibility in their behaviour patterns and this shows that they are conscious and passionate and not merely "programmed" by genetic instinct to do "this" in one situation and "that" in another situation.'[34] Don Griffin, too, based his conclusions about animal consciousness on the assumption that flexibility of behaviour was an indication of consciousness in contrast to the fixity of the behaviour of unconscious machines. He wrote, with a bit more caution: 'animals often behave in such a versatile manner that it seems much more likely than not that they are conscious'[35].

Thirty, even twenty, years ago, people could have been forgiven for believing that machines were inflexible in their behaviour. Words like 'machine-like' or 'programmed' entered the language with precisely that meaning, reinforcing the view that computers were at best 'lumbering robots'—stupid, awkward imitations of real animals, only able to do what they had been specifically told to do. But that is not an accurate description of what computers can now do.

Flexibility and the ability to learn and change and adapt to new situations are now the hallmark of modern machine intelligence. Computers can be given ears and eyes so that they can recognize objects and navigate around the world. They can learn to tell the difference between a person about to enter a car in a car park with their shopping and someone entering the same car with the intention of stealing it. And they do this not by a human specifying to them exactly what a car thief looks like (that would be too variable to capture in a simple instruction) but by letting the machine learn for itself by watching lots of instances of car theft and lots of cases of innocent shopping expeditions and thus sort out the difference. It will probably end up being much better than any human at telling the difference. Computers can be trained, learn spontaneously for themselves, and even train other computers. They have goals and can heal and tune themselves. They are the very opposite of inflexible and lumbering.

The versatility and adaptability of modern computers has thus taken away one of the key arguments that anthropomorphism has relied on up to now, namely, that if an animal is behaving 'as if' it were conscious, the simplest and most plausible conclusion is that it really is.[36] And now that those versatile, flexible computers can be made to look and behave like real animals, the 'as if' can be made even more plausible. Machines that deliberately tap into our anthropomorphic tendencies by their appearance and how they interact with us could seduce us into being very, very wrong about whether or not they are conscious. Anthropomorphism, as I have stressed all along, may be true, but it is not a reliable guide to possible consciousness in other species.

It is also striking just how easily we humans can be 'fooled', not just by complex computer programs but by the simplest ones. Even in the early days of computers there was a program called

ELIZA that set out to mimic a psychiatrist talking to a patient.[37] By picking up key words in what people typed, such as 'mother' or 'angry', and putting them into stock sentences such as 'Tell me more about your (mother/father/dreams)', or 'Tell me why you are (angry/sad/frustrated)', it could give a remarkably convincing impression of a kind, understanding person just out of reach on the other end of a computer keyboard.

So where does this leave anthropomorphism in the study of animal behaviour? Should it be avoided at all costs,[38] or might there be some, strictly limited, role for it?[39] Is it possible to keep the most important lessons from behaviorism without going to the extremes of denying animal consciousness altogether? Can ethology keep its scientific credentials and still leave open the possibility that many species besides ourselves are thinking, feeling, conscious beings?

Here are two suggestions. First, the anecdotes about animal behaviour that anthropomorphism thrives on should be taken seriously and used as possible starting points for more thorough scientific investigations. In other words, anecdotes are not evidence in themselves but they may spark off an important line of enquiry which then leads to a new understanding of what animals can do. Everyone who uses such anecdotes, however, should have a picture of Otto Pfungst and Clever Hans on their wall, in their wallet, or as their screen saver.[40] They should remember how difficult it was for Pfungst to show that the horse was *not* counting, and how he actually had to go to extreme lengths, such as deceiving the owner about what the right answer was, in order to do so. Even when he had done all that, Pfungst himself, although he knew perfectly well that Clever Hans wasn't really counting, couldn't help giving off the cue that Hans used. When the sceptical

Professor Pfungst, who had spent a lot of time and energy carrying out experiments that showed that Clever Hans wasn't really counting, stood in front of the 'counting horse', Hans was clever enough to pick up the right answer, even from him. Cleverness comes in many forms and the human form is only one of them.

Secondly, thinking anthropomorphically can be used to derive hypotheses that, if they lead to testable predictions, contribute directly to solid science. Again, the anthropomorphism does not provide evidence in itself but is a means to an end. Franz de Waal used anthropomorphism in his studies of chimpanzees to predict that chimpanzees would feel the need to 'make up' after a fight.[41] He predicted that this would lead directly to an increase in positive social interactions after aggression, which turned out to be the case. His 'reconciliation hypothesis' made completely different predictions from the previous theories of chimpanzee aggression, which were that fighting would lead to animals avoiding each other after a fight. So by using anthropomorphism and thinking what a human might want to do, de Waal gained insight into chimpanzee behaviour that it would have been difficult to get in any other way. But it is important to note that he did not stop at coming up with an anthropomorphic hypothesis. He tested it and brought it into the realm of observable behaviour, his scientific credentials intact and enhanced with a new hypothesis.

Both of these kinds of anthropomorphic thinking have important uses in a scientific study of animal behaviour, but only if they are limited to providing suggestions to be investigated by established scientific methods later on, not as evidence on their own. By using anthropomorphism in this 'subordinate' way, ethology potentially has the best of both worlds. It can use the richness of the human tendency to empathize with animals to generate ideas that might not come from any other source and it can still be scientific in its conclusions. It can stick to observable behaviour as its subject matter, but without ruling out the possibility of consciousness in animals. It can take the best from behaviorism but avoid its worst excesses.

But is 'not ruling out the possibility' the best we can do in the study of animal consciousness? The belief that many animals are conscious is the basis for many people's concern for animal welfare and so finding ways of studying animal consciousness would be a major gain for animal welfare science. The taboo on discussing thoughts and feelings in animals is now no longer as strong as it once was, so why can we not do away with it altogether and study animal consciousness openly, directly—and scientifically?

At this point, we enter controversial territory. 'Warning,' says Susan Blackmore, 'studying consciousness could change your life.'[42] She might have added: 'Extra warning—issuing warnings about studying consciousness could ruin it.' In Chapter 4, we ignore all the warning signs and look at what we really know and do not know about animal minds.

4

WHY CONSCIOUSNESS IS HARDER THAN YOU THINK

Consciousness—in ourselves or other species—has an infuriating quality. We all know what it is but it is almost impossible to define. We experience it every day but would be hard put to describe what it is like to experience anything at all beyond using a few inadequate words to describe what we are feeling—'sad', 'tormented', 'happy'. But, however many of these labels we use, there still seems to be an isolated 'I' at the centre of each of our subjective worlds.

What makes this so frustrating is that, although I can observe your behaviour, and listen to your words, I still can't know for certain what you are actually experiencing—whether your pain hurts you in the same way that mine hurts me, whether what I see as 'red' looks the same to you. Your behaviour and even what areas of your brain are active can be seen by anyone who has the right equipment, so behaviour and brain activity can be described as scientific and objective—that is, detectable from outside of you. They can be checked and double-checked by different observers. But only you can feel your feelings or live your experiences and these remain obstinately subjective—that is, private to you and known only from the inside. No one can dispute what you feel because only you have access to what you experience. That, of course, is why so many people in the past have maintained that the study of consciousness, even in humans, is unscientific and why, for much of the twentieth century, the whole subject of consciousness was avoided by most scientists.

A major step forward in the study of consciousness over the last few years has been the gradual realization that there is not just one 'problem of consciousness' but several different ones. There are what David Chalmers calls the 'easy problems', which are much more amenable to scientific investigation than what he has named 'the hard problem'.[1] 'Easy' problems are questions such as how animals and humans discriminate between stimuli or recognize objects in their environment, how they switch attention, or how they move between sleep and wakefulness. 'Easy' does not mean we now know all the answers. It means that, at least in principle, we can begin to see what the answers might be like one day. We already have computers that have eyes and ears and memories, and provide solutions to some of these easy problems. The growing subject of computer neuroscience is showing us how relatively simple nerve cells can be linked together into groups or networks capable of recognizing faces, making decisions, and even monitoring their own activity.[2] The 'easy' problems of consciousness are already part of perfectly respectable scientific disciplines with names like 'cognitive psychology' or, in the case of non-humans, 'cognitive ethology'. But the 'hard problem' of consciousness is much more difficult. It consists of a great deal more than just filling in some details on a map for which we already have a sketchy outline. It so far offers no comforting view of a future when we will understand it and bring it into the scientific fold. It seems to challenge what we think we already know and defy even a definition of what we don't know. Admittedly it is referred to as hard, not insoluble, which suggests that at least some people hold out some hope that we might understand it one day. But it is also called 'the' hard problem not 'a' hard problem because it is not one difficult problem among many others. It is uniquely baffling and uniquely, for the moment, beyond science to deal with it.

The hard problem of consciousness is, quite simply, subjective experience. Why are the things we do and the workings of our brains done consciously? Why, as David Chalmers puts it, doesn't all the information processing go on 'in the dark', free of any inner feelings?[3] The distinction between the easy and hard problems of consciousness is the distinction between giving an explanation for complex behaviour or changing brain states, on the one hand, and giving an explanation for why that behaviour or those changing brain states are accompanied by conscious experiences, on the other. The conscious experiences seem to be some sort of passenger, sitting on a bus without being part of the mechanism for propelling it. The easy problem is what makes the bus go along. The hard problem is why the passenger is there at all. We have *no idea* what the passenger is doing, despite knowing a great deal about the workings of buses.

Chalmers claims that most philosophers who write about consciousness confuse the easy and the hard problems. Typically, he argues, they start their books or papers well enough with the mystery of conscious experience, thus identifying the hard problem as the problem they are going to address. But then they begin to soften and become more optimistic. They put forward their own view of consciousness as a solution. However, on closer examination, the 'solution' turns out to be a solution to one of the easy problems, not the really hard problem at all. The same can be said about biologists (including, in the past, me). Discussions of animal consciousness acknowledge the 'mystery' of consciousness but then proceed to argue that because we have begun to understand some of the easy problems such as concept formation, learning, flexibility of behaviour, wakefulness, attention, and so on, we have, unless you want to be fussy to the point of being pedantic, solved the hard problem as well. By a sort of sleight of hand, what starts out as solutions to easy problems becomes evidence for conscious experience itself. We are asked to believe that the 'explanatory gap' between the physiology and behaviour we can observe and the subjective experience we cannot is no more than a small ditch that we can hop across.[4] We don't need bridges across the explanatory gap any more because the distinction between the 'easy' and 'hard' problems has all but disappeared. Hop across and we can all stop arguing.

This chapter is about the fallacy of this argument. It is about the hardness of the hard problem and the continuing, infuriating (but not necessarily totally long-term intractable) nature of consciousness in ourselves and other species. Above all, it shows why we should not rely on solving the hard problem to make the case for animal welfare.

Now, I am aware that there may be many people who believe that questioning or facing up to consciousness in this way is a complete waste of time or, even worse, will actually hinder progress in animal welfare.[5] Some people will feel that the most reasonable starting point is to assume that many animals are conscious and that the onus should be on sceptics to show that animals are not, rather than the other way round. Others will go even further and say we should all burst out of our behaviourist chains and embrace consciousness in animals as an undoubted fact. Such arguments—and they come from scientists as well as non-scientists—simply ignore the hard problem altogether or assume that it has been solved. But in doing this, they also betray a lack of understanding of what the hard problem really is. Failing to realize how far we are from solving it is not just a minor error. It could be very damaging to animal welfare in the long run.

If animal welfare science gets a reputation for claiming as scientific fact things that cannot be studied by science then genuinely scientific claims will be dismissed as 'just animal welfare science' (or, even worse, the claims of 'welfairies'). People who are not at the moment at all concerned about animal welfare will be *less* likely to take it seriously if they can show that the case for animal welfare is based on insubstantial evidence. By confusing what is scientifically testable with what is not, even the solidly based evidence will lose some of its impact. The only way to deal with the hard problem is to respect it. We do not as yet have a way of solving it and we should not pretend that we have. For the foreseeable future, the hard problem remains very hard indeed.

So what exactly is so hard about consciousness? What is it that makes it so uniquely difficult to explain? And why should the 'explanatory gap' between what we know and what we would like to know be so much more treacherous to cross for consciousness than for any other problem for which we don't yet have a solution? Part of the answer can be glimpsed by considering something that we know is definitely not conscious—a security camera.

Security cameras are sensitive to movement and respond appropriately by switching on a light, sounding an alarm, or even ringing a police station, but most of us don't worry too much about whether or not they are conscious, despite our tendency to describe them in anthropomorphic terms ('Don't do that or the camera will think you are an intruder').

We know that what a surveillance camera does is very simple. It can detect movement and it can then respond in a totally automated way to raise the alarm and even to summon the police. We also know that if we looked out of the window and saw a strange man running across the lawn at night brandishing a gun, we would perform a similar task of alerting the police but we would do it in a completely different, conscious way. The end result is the same, but with a different way of getting there. One is the totally unconscious activation of a phone line, the other has the full panoply of conscious recognition of the presence of an intruder, followed by the experience of fear at what he might do, and then the conscious action of telephoning the police and explaining to them rationally what is happening.

This simple example shows why identifying where there is consciousness is so difficult. There is clearly a spectrum of mechanisms for producing a similar outcome that has security cameras at one end and ourselves peering into the night at the other. Where on this spectrum are we to put, say, slugs? Fish? Chimpanzees? Plants? The fact that so many of the attributes of consciousness, such as the ability to respond to stimuli and choose an appropriate action, can be mimicked by relatively simple machines shows that it is not necessary to feel or experience anything in order to have adaptive, appropriate behaviour. A few simple sensors, a bit of programming, and an electrically powered output of a sort we are all familiar with and you can do a lot of routine, everyday behaviour. Consciousness just isn't necessary.

An obvious objection is that we, along with most animals, are much more complex than a camera fixed to an alarm system. Consciousness might not be necessary for such a basic mechanism that does nothing more than detect movement, but we are not cameras hitched to phone lines and neither are other animals. Even a garden snail, with about a million nerve cells, way surpasses such a crude system. In more complex animals, let alone us, consciousness might well be much more likely.

This argument—that consciousness isn't necessary for simple mechanisms but 'kicks in' at some point and becomes necessary for more complex ones—is superficially very attractive but is itself problematic. It only postpones asking the hard question rather than answering it. At what point does consciousness become necessary? After all, we now have security cameras that can recognize faces and do much more than just respond to movement. So, how complex must something be before it 'has' to be conscious? And how does being conscious help make the animal or person (or machine) more effective?

Donald Griffin, who did so much to persuade people to start thinking about animal consciousness, stressed the flexibility and adaptability of much animal behaviour as a prime reason for thinking that many animals are conscious. An example of the kind of flexibility that impressed him was one he saw with his own eyes.[6] He describes how he watched a group of lionesses hunting together in a way that seemed to him to be tactical and planned, rather than instinctive. Instead of all rushing at a wildebeest at once, the lionesses split up, one of them moving away from the main group into a position where she was hidden from the wildebeest. The others then drove the prey towards their hidden companion, which then caught it. Griffin argued that we should at least consider the possibility that the lionesses knew what they were doing and planned the whole ambush. He was sure that what he had witnessed was the opposite of rigidly fixed hunting behaviour. The lionesses were adapting their behaviour to the particular terrain and coordinating their behaviour with what the others in the group were doing.

He also saw the same flexibility and adaptability in the way beavers build dams, and even more so in their ability to deal with damage to their dams. If a beaver dam is damaged and water starts leaking from it, the response of the beaver is flexible enough to deal with that particular leak. Beavers may have to take very different courses of action depending on where the dam is, where the water is coming through, and what material is available to repair it. The fact that the beavers can not only locate the source of the leak but then repair it suggested to him that the animals understood the connection between their construction and its ability to hold water. Ingenuity and the ability to deal with novel situations he saw as clear evidence of conscious awareness because it put animals in a different league from simple machines, preprogrammed to do just one or two simple tasks. Animals did not act just out of instinct or simple trial and error learning. Animals, like humans, could look at a completely new situation, one where neither genes nor previous experience could tell them what to do, and work out a solution *in their heads*. Animals could think.

There is now a substantial body of evidence in support of the idea that animals can apparently think their way out of difficult problems. This includes claims that chimpanzees can work out what is going on in the mind of another chimpanzee and birds that can understand how to use novel tools to get their food.[7] Much of the evidence is still controversial but the field of 'animal cognition' is now well established, scientifically respectable, and regularly gets write-ups in the popular press. Thinking animals, with the implicit assumption that 'thought' means 'conscious thought', catch the public eye and enter the public imagination.

However, they have also attracted an increasing number of what Dan Dennett has called 'killjoy' explanations.[8] A killjoy explanation is one that takes behaviour that seems to imply human-like thought but in fact can be explained by much simpler explanations such as innate behaviour or learning.[9] For example, perhaps the beavers have a repertoire of standard responses to leaks and run through their repertoire until something works. By a combination of innate responses and learning, they repair the dam, giving the appearance that they knew all along what to do. In reality, they operate a 'try everything and see what works' approach, which often gives a remarkably good result. An experienced beaver would then be able to repeat the successful strategy the next time there was a leak, further reinforcing the view that it understands the physics of dam construction and water retention. There would be a sense in which the beaver did understand the situation but the understanding would be limited to connecting a particular behaviour to a particular outcome; no more remarkable than the way that a rat understands that pressing a lever gives it a pellet of food, or gulls understand that fishing boats are a source of food.

Similar 'killjoy' explanations have been given for a whole range of other examples of animals behaving in adaptive flexible ways and appearing to think about what they are doing. The idea that chimpanzees might have a Theory of Mind and be able to deliberately deceive each other was initially widely accepted, but has now been challenged on the grounds that all chimpanzees do is simply watch the body language of the other animal very closely.[10] The Theory of Mind hypothesis implies that chimpanzees are capable of actually working out what another animal might be thinking by putting themselves 'in their shoes'. So a chimpanzee might use this knowledge of other minds to deliberately deceive another animal into thinking something that was false by walking away from a source of food rather than towards it. If chimpanzees have a Theory of Mind, they will be thinking something to the effect: 'I don't want that other chimpanzee to think that I know where the food is so I shall make him think that I don't know by walking in the opposite direction until his back is turned.' The killjoy explanation is that the chimpanzee has learnt by bitter experience that if he goes straight to food when another animal is following him, the food will be taken away. If on one occasion he happens to walk in the opposite direction for some reason, he might have the pleasant experience of not being attacked and of eventually keeping the food for himself. All he has to do is to watch the other animal to see if it is following and adjust his behaviour accordingly. He understands that the best thing to do if he is followed is to move away from, rather than towards, the food, but he discovers this by accident and then realizes what a good move it is. He did not plan ahead to think how he could possibly plant a false idea into the mind of another animal.

The 'killjoy' and the 'working it out in the head' theories can be distinguished by observing the history of what happens when animals are first introduced to the situation where the subordinate has its food stolen by a dominant. The 'killjoy' explanation predicts that the subordinate should initially spend time losing food to the dominant but then, by chance, on one occasion do something that prevented or at least delayed the food being stolen. He should then repeat the accidental solution because it resulted in him being able to eat the food, possibly modifying it as even better solutions are discovered. The 'working it out in the head' hypothesis, however, needs no such happy accident. The subordinate should put the right solution into practice straight away, having imagined what would happen beforehand.

These predictions were tested by some careful observations of interactions between a dominant male (known as Boss) and a subordinate female, called Rapide, in a group of highly sociable black monkeys called sooty mangabeys.[11] A competitive food situation was set up between the two by hiding food and then letting both monkeys search for it. But the subordinate Rapide was given an advantage. She was always shown where the food was hidden before being let into the arena where the food was, while the dominant, Boss, was only shown sometimes. On the occasions when Boss knew where the food was ahead of time, he went straight to it. But on the occasions when did not know where the food was, he followed Rapide, waited until she found the food and then simply took it from her. Rapide quickly got the idea that if Boss was following her, her best bet was to move away from the food, not towards it. She quickly developed the ability to adjust her behaviour according to whether Boss did or did not know where the food was and only adopted the 'deceptive' tactic when he did not know.

Seen when Rapide was experienced in her dealings with Boss, this looks as though Rapide was deliberately deceiving him and working out what to do by what she thought he knew. But a closer look at the history of the interaction makes the 'killjoy' explanation much more likely than deliberate thinking deception. On the first few occasions, the naive Rapide went straight to the food and immediately had it taken away from her by Boss. Then on one trial, she was distracted by something and failed to go straight to the food. Boss did not follow her and when she did go and find it a few minutes later, Boss's back was turned and she was able to eat in peace. She had accidentally discovered that, if she were being followed, walking past the food and waiting a few minutes before going to it was a much better way of getting something to eat than going straight for it. She subsequently repeated this happy accident, but only when Boss did not know where the food was. She did not need to have any idea of whether or not he had seen the food being hidden. She just adjusted her behaviour depending on whether or not he was following her.

So, serendipity is not just the prerogative of us humans. Animals of all sorts have an extraordinary ability to notice connections between events that might initially happen by chance and then take advantage of them. They behave in ways that look like insight into the future, planned and thought about. In fact it is more often a well-developed capacity to notice what has happened in the past and capitalize on the connection between one event and another.

But even if there are killjoy explanations for insight and mind reading, and being able to plan for the future, we can still ask what exactly the nature of the joy is that is being killed. Has it killed off the easy problems of consciousness or has it actually dealt a blow to the hard problem? Is it the joy of believing that the behaviour had a complex explanation that now has to be replaced by the 'disappointment' that it could be explained in simpler terms? Or is it the joy of believing that the behaviour, now that it is explained in simpler terms, no longer indicates consciousness? These are two different sorts of joy and they are not necessarily killed by the same sort of evidence.

The first kind of joy-killing should not really be called joy-killing at all. It involves nothing more than substituting one explanation of an 'easy' problem for another explanation of the same easy problem. There may be some regret at the loss of a 'pet' hypothesis that doesn't stand up to test, but in the grand scheme of things, suggesting and then disproving ideas is the way that science advances. Even theories that eventually prove to be wrong are part of progress because they stimulate thought and they make people find out more about the world. So to give a plausible explanation of how beavers build dams or one monkey interacts with another can hardly be described as 'killjoy' just because it replaces one 'easy problem' explanation with another. The beavers and the mangabeys could still be conscious even if their behaviour now has a different explanation.

But can the 'joy-killing' really be of the second kind? Can it actually kill off the joy of believing that animals are conscious? That would imply that the hard problem has already been solved so conclusively that we are now in a position to say exactly when consciousness can be ruled in and when it can be ruled out. But the hard problem has not been solved. Until it has, until we have a coherent theory of what consciousness is, we have no more reason to believe that clever animals are conscious than that stupid animals, acting out of 'blind instinct' or trial and error learning are. The hard problem does not even flinch in the face of a bit of joy-killing.

As we still have no idea about what consciousness is, we have no grounds for claiming that consciousness only springs into existence when animals show evidence of a certain (undefined) capacity that has some (unexplained) ability to solve a difficult puzzle. Why should we assume that innate responses or trial and error learning are not also done consciously? Flexibility and adaptability are the hallmarks of much animal behaviour, such as nest-building, song learning, tool use, hunting and finding food. Most of these come about during development as a mixture of innate responses that are then modified by learning so that although an animal may be 'programmed' by its genes to respond to certain stimuli, it then learns to modify its behaviour as a result of experience.[12] Genes are a lot more flexible than their 'one-gene-one-behaviour' reputation gives them credit for, so that the combination of innate behaviour and learning gives animals huge repertoires of adaptive and flexible behaviour, exactly the characteristics that thinking is supposed to confer. So if thinking animals are conscious, there seems to be no obvious reason for ruling out consciousness for adaptiveness and flexibility achieved in these different ways. Innate behaviour and learning may be signs of consciousness too.

Paradoxically, by accepting the hardness of the hard problem and acknowledging how little we really understand about what it is to have conscious experiences, we leave open the possibility that many different animals, right across the animal kingdom, might have such experiences. It is those who think the hard problem has already been solved by pinning consciousness to a certain kind of complex behaviour who are the ones who rule them out. Respecters of the hard problem—consciousness sceptics—leave far more species still in the running for membership of the consciousness club.

For example, some people believe that the hard problem has been solved because it is associated specifically with what are called Higher Order Thoughts (or HOTs).[13] Higher Order Thoughts are thoughts about thoughts; for example, having a thought about something and then thinking that thought might be wrong. You might have a plan about how you will organize the food for a party next Thursday and then review what you had planned and realize that if carried through, the plan would be disastrous because you would be way over budget. Your lower order thoughts (about the food you would serve) were monitored and in this case vetoed by the higher order thought that your plan was a bad one. Higher Order Thoughts are believed by many people to be associated with language because it is difficult to imagine having such HOTs (about, for example, what is going to happen next Thursday) without language to frame them in.

Some people go so far as to argue that because animals do not have language they cannot have HOTs and are therefore not conscious.[14] It is certainly a fascinating issue in animal behaviour to try to discover whether animals are able to monitor their own thoughts and whether they can plan ahead. But even if they could, there is no real evidence that having HOTs is associated with consciousness, and not having them is not. It is just a belief, like any other belief, of where consciousness kicks in. It throws a hammer at the hard problem, there is a loud clang, but the hard problem is still there. We can use a study on rhesus monkeys to see why even behaviour that appears to suggest that an animal is having thoughts about thoughts does not guarantee consciousness.

Robert Hampton wanted to see if monkeys could monitor the state of their own memories,[15] a first step in seeing whether they could have higher-level thoughts. He trained two monkeys on a task called 'matching to sample'. The monkeys would be shown a sample image, such as a picture of a chicken, and then a short time later, given a choice between that picture and three other pictures shown on a touch screen. The monkey had to look at the four pictures on the touch screen, remember which one he had been shown before as the 'sample', and then pick this one out. If he touched the correct image, he was given a peanut as a reward. Monkeys like peanuts and these two readily learnt to do this task.

However, Hampton was also, during the same period, training his monkeys on a much easier task, one in which they still saw a sample picture but they did not have to remember it. What they saw on the touch screen was a single image and all they had to do was to touch it and they got their reward. The catch was that with this simple task, the reward was not a peanut. It was a piece of monkey chow. Monkeys like monkey chow but not as much as they like peanuts. So in the more difficult trials, the monkeys were given a kind of food they really liked (a peanut), but they had to remember what the sample picture was and make the correct choice between four pictures. If they could not remember and chose the wrong picture, they got nothing at all. In the easier trials, the monkeys did not have to remember anything, and were given food every time they touched the screen, but the food was not the sort they liked best.

Hampton gave the monkeys these two sorts of trial interspersed with each other. Sometimes he, as the experimenter, decided which sort of trial the monkeys were going to have—difficult or easy. But sometimes he let the monkeys decide for themselves whether they wanted to try the difficult task with the favoured food and risk getting nothing if they got it wrong, or stick with the easy 'safe' task with the boring food. He wanted to know if the monkeys would choose the easy or difficult task depending on how good they thought their own memories were, in other words, if they were monitoring their own thoughts. He reasoned that if the monkeys were capable of monitoring the state of their own memories (if they knew how well they had remembered the sample picture, in other words), they should be able to decide which sort of trial to go for by how well they had remembered the sample. If they had a very clear memory of what the sample was, they should go for the more difficult trial because they would know which image would give them the highly desirable peanut. But if they had forgotten what the sample looked like, they would not risk getting nothing at all in the difficult trial and instead would be better off going for the certain, if less desirable, monkey chow they knew they could get from the easy, no memory trial.

On the other hand, if he, the experimenter, decided which trial the monkeys should have, he would have no way of knowing how well the monkeys had remembered the sample image on a given trial. So when he chose to give the monkeys a difficult trial, they should do less well than when the monkeys chose for themselves to go for the difficult option because only they would know how good their memories were.

The results suggested that the monkeys were indeed behaving differently depending on the state of their own memories.

They were more accurate in the difficult trials when they were allowed to choose for themselves whether to go for the difficult option than when Hampton decided for them. They appeared to know more about the state of their own memories than he did.

On the face of it, this looks as though the monkeys were able to monitor their own thoughts and so were fulfilling one of the key criteria for the HOTs theory of consciousness. But wait a minute. It is very easy to mimic such behaviour on a computer. Imagine a simple machine programmed to choose the brighter of two lights, A and B. It measures the brightness of each one with a light meter, such as would be found in any camera, and remembers the two values in lux. It then subtracts one from the other to give a third value, C. It 'chooses' whichever of A or B is the larger but it also uses the value of C to give it a measure of how certain it is about its choice based on the size of the difference. For example, if the difference between A and B were very large, it could be programmed to say: 'I prefer A and I am very certain that this is the brighter light'. If the difference were very small, it would say 'I am choosing A but I am not certain I am right'. Add in a bit of hesitation for the uncertain choice and you would have a machine that behaved uncannily like a monkey or even a human being making a more difficult or less difficult choice and apparently 'thinking about its thoughts'. This isn't just an easy bit of programming to do. It's a trivial bit of programming. The machine remembers one single number (the difference in brightness between the two lights) and that is all it needs to appear to be engaging in Higher Level Thoughts. The monkeys in Hampton's experiments could similarly have been using the single value (such as the strength of a memory trace that decayed over time) and have learnt to choose difficult or easy tasks depending on whether this value was above or below a certain threshold. This would no more need to be done consciously than a car needs to be conscious to monitor the amount of fuel in its own fuel tank.

Monitoring the strength of a memory may be done consciously but it doesn't have to be. We have no more idea of whether it actually is done consciously by monkeys than we know whether the antelope fleeing from a pride of lions consciously feels fear or whether the lions have consciously worked out where the antelope is likely to run. Or whether the beaver plugging a leak in its dam is conscious of how effectively the next branch is going to stem the flow of water. Without a much better theory of consciousness than science can offer us at the moment, we have no reason to associate consciousness exclusively with complex 'cognitive' examples of animal behaviour or with particular kinds of thinking.

So, on behalf of all the not so clever animals that achieve an impressive amount with the simplest of mechanisms, it needs to be clearly stated: 'killjoy' explanations may show that they aren't thinking as we understand it, but they do not show a lack of consciousness. They give us alternative explanations for some of the easier problems of consciousness, such as how decisions are made in the face of changing amounts of information. But killjoy explanations don't kill off consciousness itself because we don't have sufficiently coherent solutions to the hard problem for any of them to be killed off. Killjoy explanations actually keep the possibility of conscious experiences alive in many other species, even the stupid ones, because they bring us back to the real issue. This is that there is a huge explanatory gap between what we objectively know about brains and what we subjectively don't understand about conscious awareness and therefore we have no idea which species have brains that are conscious. Facing up to that uncomfortable fact is, in the long run, going to help the case for animal welfare, far more than pretending to have evidence of which animals are conscious and which ones are not, when in fact nobody knows. And it is going to include far more species than ruling some of them out just because they don't show the kind of behaviour or brains that might, but only might, be harbouring a little flame that we would recognize as conscious experience.

What does facing up to these issues actually mean in practice? It means making a very difficult choice. You may be getting a little tired by now of my geographical metaphors—roads and gaps and bridges—but I'm afraid another one is coming up. Once we acknowledge the hardness of the hard problem, we arrive at a very definite crossroads in thinking about consciousness.

Pointing in one direction is a sign that says 'Physicalism: strictly biological explanations of consciousness'. This does not look like a particularly easy road. The map says simply that consciousness arises entirely from the activity of brains, but does not say exactly how. Pointing in the other direction is a sign that says 'Dualism'. Dualism, says the map, is the idea that it is impossible to give a complete explanation of consciousness just by understanding the workings of the brain because something would be left out. That something is consciousness. However much we learn about how the brain works, we can never explain consciousness because it is not explicable in physical terms. It is an extra, something mysterious and non-material. It is the Ghost in the Machine. This sounds more promising, but what does it look like when you get a bit further down the road?

Dualism gets its name from the fact that it sees the working of our brains as a kind of duet between two kinds of stuff—the physical 'stuff' of the body (including the brain) and the mental 'stuff' of the mind, a theory put forward by the French mathematician and philosopher René Descartes in the seventeenth century. Descartes and the many modern-day Dualists who still follow his lead, believe that minds cannot be explained in physical terms. However advanced science becomes, however detailed its knowledge of brain cells and their interconnections, consciousness will still not be explained because it is due to something else.

Dualism runs into trouble because of the vague nature of this something else or 'mind stuff' that is supposed to explain consciousness, and because it insists that the mind stuff influences the brain by a kind of magical interaction, modifying nerve cells but not through any physical means. It cannot explain how the mind stuff (not subject to physical laws) pokes its fingers into the workings of brains that are subject to physical laws and somehow makes them behave differently. It is a bit like trying to explain everything about how the engine of a car works and then saying there was a little gap—a missing flywheel or belt—that wasn't physical at all and was filled with a mysterious magical substance that didn't follow any known physical laws but somehow managed to control the working of the whole engine. To most people, that would sound completely absurd. An engine with gaps wouldn't work. Everything has to be linked together in ways that are understandable by the laws of physics and chemistry or we don't have a workable engine.

Using 'consciousness' as part of the mechanism underlying the behaviour of animals or people could be equally absurd, at least if 'consciousness' is taken to mean something other than the workings of the brain. If conscious experiences are seen as more than just the physiological activity of our brains, something magical and extra that cannot be explained by biology, physics, or chemistry, then we have an engine with a gap. It asks us to believe that the laws of physics and chemistry hold all over the solar system, in the fuelling of stars and the movement of planets, in the tides of the earth and the workings of our bodies below the neck, except in one place—inside our heads. It asks us to believe that the laws of physics and chemistry are true everywhere else except as explanations of consciousness, when we have to invoke a completely different kind of stuff, as yet unknown to science, that magically interacts with the rest of the known universe and has the power to influence it profoundly. Strong magic.

The problems with Dualism—its reliance on mystery and magic—has led most scientists to reject the Dualist road, seeing it as leading to a kind of intellectual honeytrap—superficially attractive but fatally flawed. Instead, they turn down the other road, towards Physicalism and the idea that consciousness has a completely biological explanation and needs nothing more to explain it than the workings of the nervous system. The brain, like everything else we know, obeys the same physical laws and is made from the same physical stuff. It is a machine, even though a very complex one that we do not yet fully understand, but there is no ghost in it. The brain, with its billions of neurons and the billions of connections between them, is all there is, so that if we fully understand the brain, then we would understand consciousness, which is a property of certain kinds of neural processing. The fact that we cannot see consciousness when we look inside a brain is no more puzzling than the fact that we cannot see a computer game if we look inside the computer running it. Consciousness is like a supremely clever piece of software that we are not yet able to decipher.

The choice at the fork in the road is a stark one. Is the physical brain all there is, with consciousness arising from certain kinds of neural processing we don't yet understand? Or is there magic in there?

Most scientists don't like to believe in magic and so ought to be rejecting Dualism utterly. But, very confusingly, many people, including many scientists, are secret Dualists. Or perhaps it would be more accurate to say that they are unwitting Dualists. They don't mean to be and they don't think they are, but when the going gets tough, they weaken and start talking about consciousness in ways that have a distinct whiff of Dualism about them. To help you decide which road to take, and to discover whether you are really a secret Dualist, here is a test in the form of a thought experiment, known as the Philosophers' Zombie.[16]

The test involves imagining a zombie. A zombie looks like a human being and behaves exactly like a human being, including interacting with them socially. The body of the zombie is made of flesh and blood, and it has a brain that, if put into a brain scanner or subjected to any of the tests that might be made on a human brain, would give exactly the same results as a real human brain. If you were to pinch the zombie, it would say 'ouch' and appear to get annoyed in exactly the same way a human being would. The only difference between a zombie and a real human is that the zombie is not conscious. It has no subjective awareness of anything at all. The key question is: do you believe that a zombie could exist?

If you answer 'yes' to this question, then you are, whether you realize it or not, a Dualist. If you believe that it would be possible for a zombie to have exactly the same brain circuits and exactly the same brain activity as a conscious human and yet not be conscious, you must also believe that consciousness is more than just the result of what brains do. You would be claiming that the zombie lacks the magic, the mental 'stuff' for consciousness.

Only if you answered 'no' to the question of whether a zombie could exist are you a squeaky clean non-Dualist Physicalist. For you, a zombie could not exist because if it had the same brain states as a conscious human being, it would be a conscious human. Those brain states would make it conscious and it would cease to be a zombie. We may not be able to identify what those critical zombie to non-zombie brain states are at present, but that is simply because we do not yet know enough about the way brains work.

This thought experiment makes it clear that there are 'mysteries' about consciousness down whichever road we choose to go but those mysteries are of two quite different kinds. The first kind of mystery is the Dualist mystery. Consciousness is mysterious because it is made up of a magical kind of mental stuff that does not obey physical laws and is different from anything else in the known universe.

The second kind of mystery is the non-Dualist scientific mystery, the mystery that Physicalists still face. Consciousness is mysterious because it arises when certain kinds of brains are in certain kinds of states and we cannot as yet identify the crucial difference between states when consciousness is there and states when it is not. We may not be able to explain this difference with our current state of knowledge, but conscious brains are believed to obey the same laws of physics and chemistry as unconscious ones and as the rest of the universe.

Both routes can be said to be mysterious. But we take a step forward if we make it clear whether we mean 'mysterious' in the Dualist sense of magic, and the need to invoke something extra to the workings of the brain, or whether we mean 'mysterious' in the Physicalist sense that the brain is all there is and we just don't know yet how the working of the brain gives rise to consciousness. It sounds like a small distinction but it has a profound effect on how we approach the problem of consciousness in both other animals and ourselves.

We can see just how profound this distinction is by taking a question that biologists frequently ask about consciousness: what is the function or adaptive significance of consciousness? In other words, how does it help survival? Why should natural selection favour the evolution of conscious beings over non-conscious ones?[17] This seems a very natural question, given that biologists constantly ask similar questions about the evolution of every other aspect of animal and plant life, such as why they are the shape they are, why they live in groups, and so on.

But the question becomes subtly different depending on whether you have adopted a Dualist or a non-Dualist position. To the Dualist, it means why did humans and animals evolve conscious experiences over and above their complex workings as machines? They aren't 'just machines'. They have this extra bit of conscious awareness that no mere machine has. So what is it doing? If it evolved by natural selection, it must be doing something, making a difference so that conscious animals survive and reproduce better than non-conscious ones. The search is on for the function of consciousness as the elusive plus factor.

To a strict Physicalist, on the other hand (and there are fewer of us about than you might think), it simply doesn't make sense to ask what the function or adaptive value of consciousness is, independently of the adaptive value of the neural processing in the brain that gives rise to it. It is not hard to see what advantages were gained by organisms that could do certain complex behaviours—such as to make plans for the future. But if these conscious processes are simply the result of certain kinds of neural processing, then all that natural selection has done is to act on the neural processing. It won't act separately on the consciousness that is associated with that neural processing. Consciousness has its evolutionary advantage because of the advantage of its underlying neural processing, not because of anything additional. So we should not ask why natural selection favoured consciousness but rather, why natural selection favoured the particular kind of neural processing that is associated with consciousness. It's the neural processing that is exposed to natural selection and makes the difference between a conscious and a non-conscious being. To ask why natural selection favours consciousness independently of neural processing is to believe in zombies and Dualism and magic.

Nowhere does the distinction between Dualism and Physicalism pose quite such a challenge as with that most quintessentially conscious experience of all, the experience of being in pain.[18] What is the function of pain? On one level the answer is obvious. Animals (and ourselves) have clearly been evolved by natural selection to have fast, efficient, and reliable ways of avoiding situations that are damaging. To this end there are sensors in the skin, special emergency neural pathways that give top priority to damaging events such as burning or cutting of the surface. It is these 'tissue damage detection circuits' that natural selection has been working on. So why, apparently in addition to what the neural or electronic circuits are actually doing by way of implementing avoidance behaviour, does the pain have to *hurt*? What does the conscious experience of pain add to an organism's ability to survive that unconscious neural pathways couldn't provide?

Answering this question by saying we know what the value of pain is because we know the damage that happens to people who cannot feel pain, is, inadvertently, a Dualistic explanation of the function of pain. People who are unable to feel pain do indeed damage themselves, often fatally.[19] Without feelings of pain, they don't notice when their hands have been cut or their bones broken. There is clearly a malfunction of the nervous system somewhere between the peripheral nerves that detect the tissue damage, the brain that receives the information, and the muscles that move the injured limb. But there are two results of this malfunctioning. One is that the nervous system is damaged and fails, like a broken engine, to do its job of moving the limb properly in response to damage. The other is that there is no experience of pain. To ask about the function of the experience of pain as though it could be separated from its underlying neural platform is Dualism.

The Physicalist view is that the conscious experience of pain is entirely due to the activation of certain neural circuits and brain regions, and so people who are unable to feel pain must also have something wrong with their nervous systems somewhere, even if it is not obvious what it is.[20] They couldn't have fully functional nervous systems, doing exactly the same as those of normal people able to feel pain, and not feel pain unless the experience of feeling pain were a Dualist extra, something separate from the physical working of the body. The Physicalist view of the function of pain (the experience) thus cannot be separated from the function of having a nervous system that is particularly good at avoiding damage. Natural selection acted on the neural pathways and some of those neural pathways—for reasons we have yet to understand—supported the experience of pain. Pain, like other experiences, is a property of neural circuitry and cannot be separated from it. People who can't feel pain have the wrong sort of circuitry.

Dualism or Physicalism, we are left without answers to the very questions we most want to know the answers to when we turn to animal consciousness. If consciousness is nothing more than what happens when particular sorts of neural processing take place, what exactly are these special sorts of neural processing? Why can't the neural circuitry just get on with its processing in a purely non-conscious way? Why does it have to drag consciousness in at all if it doesn't do anything extra? And when, in the course of evolution, did this consciousness-bearing kind of neural processing first creep in to the nervous systems of other species? Time and time again we have come up against the hard problem and its infuriating ability to frustrate all our efforts to find answers to such questions. Perhaps it is time for an assault from a different direction. Perhaps we need to start with the one thing we think we know about consciousness with some certainty—that we ourselves are conscious—and look at the way scientists are now beginning to study human consciousness. Perhaps the study of human consciousness is the key to what happens in other species.

5

CONSCIOUSNESS
UNEXPLAINED

The study of human consciousness is now, in the twenty-first century, flourishing as never before. There are literally hundreds of books on the subject as well as whole journals entirely devoted to publishing scientific papers on consciousness. Consciousness in humans is not just scientifically respectable. It is at the cutting edge of brain research.

So how has the study of human consciousness managed to achieve its newly found respectability? How does it escape the behaviourist criticism that consciousness is outside science? What has happened to change the minds of human psychologists and convince them that consciousness can now be studied scientifically, despite obvious objections that consciousness is subjective and we can never 'really' know what another person is experiencing? And if this can happen for human psychology and human conscious experiences, why not for animal behaviour and the study of animal consciousness?

One development that appears to have played a major part in turning the study of human consciousness from unscientific to scientific is new technology that allows us to obtain images of living, thinking brains. Modern brain-scanning techniques, such as fMRI (functional magnetic resonance imaging) now make it possible to look inside the skulls of people (and other animals) to see which parts of their brains are active when they are doing different things, such as smelling a rose, touching a hot object, or making a decision. What is actually being detected is the consumption of oxygen from the blood in different parts of the brain,[1] but since brain cells use oxygen to pass messages to each other, their oxygen consumption is a sure sign of their activity. Different levels of this oxygen use are then plotted on a map of the brain with different colours so that it is possible to watch the shifting patterns of colour on these maps as different parts of the brain become active and 'light up' with a new colour. What the maps show are thus not thoughts or feelings or decisions in themselves, but the telltale signs of parts of the brain that are working a bit harder and having to use more oxygen.

The hope, when these techniques were developed, was that it would be possible to identify the parts of the brain that were active when people were conscious and inactive when they were not— the 'seat' of consciousness, as it were, or, as psychologists called it, the NCC (Neural Correlates of Conscious).[2] This was a clever move. The stated aim was not to make yet another fruitless attempt to solve the hard problem by trying to explain how the brain caused or gave rise to consciousness. The much more limited aim was to take the first, tentative steps towards such a solution by identifying what activity in the brain was correlated or went along with consciousness. That would still leave open the much more difficult issue of how the parts of the brain that were associated with a person being conscious actually made them conscious, but it would tell us the best places to look to find out. Ideally, the NCC would turn out to be an identifiable 'nugget' of nervous tissue that was active or inactive depending on whether a person was conscious or not.

Such a quest for the location of consciousness would have had enormous consequences for the study of consciousness in non-human animals.[3] Having identified the human correlates of consciousness, it would then be possible to turn to other species one by one and ask whether they, too, had similar brain activity. If the brains of other animals showed brain activity similar to that shown by humans when they report being conscious of something, then the analogy between them and us would be very close indeed. We wouldn't even need words. We could use brains. If other animals had human-like 'nuggets' in their brains, we could do a comparative study of consciousness in different species. Based on the size of their nuggets, we could decide whether they were conscious like us, a bit conscious, or not conscious at all.

Initially it seemed that the search for human consciousness would be quite straightforward. There are a variety of conditions when people drift from unconsciousness to consciousness, such as being asleep or awake, or coming round from an anaesthetic, and these seemed ideal opportunities to study their brains to see when and where consciousness was present.[4] It would be necessary to ask people at first at what point they became aware of different things but then, once having calibrated what they said with what their brains were doing, this more objective evidence of brain activity could be used for future studies and then extended to people who were unable to use words because of brain damage—and finally even to non-human animals.

Unfortunately, human consciousness has turned out to be much more elusive than this. The one thing that emerges from the very large number of brain imaging studies that have now been done is that there is no simple connection between being conscious and what a brain is doing.[5] Take anaesthesia, for example, where you would have thought there would be a clear difference between a conscious brain and an unconscious one. Anaesthesia can be induced in a controlled way by drugs and many of us have experienced what it is like to 'go out like a light' when the anaesthetic takes hold. Looking at how the brain is responding when this happens, and again when someone comes round as the drug wears off, would seem to offer the real possibility of pinpointing what happens when consciousness is lost and then regained.

In fact, general anaesthetics do not knock out a specific part of our brains. Instead, they affect the transmission of information all over the brain, so that large numbers of cells communicate with their neighbours less effectively. What appears to us to be a very sudden change in our conscious state from awake to going under is in fact achieved by a very widespread, generalized change in many parts of our brain all at once. Anaesthesia has turned out to be much too blunt a tool for pinpointing consciousness.[6]

Looking around for a more precise way of switching consciousness on and off in a particular part of the brain, psychologists have come up with a very ingenious way of doing this (although it should be noted that they are still reliant on asking people what they are experiencing for knowing when the switch happens). They make use of subliminal stimuli. If a picture of, say, a person making an angry or fearful expression is flashed up in front of someone for a very short time (less than 40 thousandths of a second or 40 ms), then they almost never consciously see it and, if asked, usually report that there is no face there at all. But if the same face is flashed up for a bit longer (more than 170 ms), then they report seeing it. So it is possible to turn conscious perception of the same stimulus on or off simply by controlling how long people see it for and this has given a very precise way of discovering which brain activity is associated with reports of consciousness.

Using brain imaging (fMRI), an international team based in London showed pictures of fearful or disgusted human faces to their volunteer subjects for different lengths of time and looked at the activity of their brains.[7] They found that when the faces were presented for long enough for the subjects to report that they saw a face (i.e. over 170 ms), then a part of their brain called the amygdala was very active if the face showed a fearful expression, whereas another part called the insula was silent. However, when the face showed an expression of disgust, there was more brain activity in the insula than in the amygdala. This clearly showed that different parts of the brain were 'interested' in different facial expressions and that this was associated with people being able to report on what sort of face they had seen.

But everything changed when the same faces were presented subliminally—that is for such a short time (less than 40 ms) that the subjects reported seeing no faces at all. Now the difference between the two parts of the brain disappeared. The amygdala did not respond to the fearful faces and the insula had no response to the disgusted ones. There appeared to be a clear-cut difference in the activity of the subjects' brains when they were conscious of seeing the faces (or at least reported seeing them) and when they were not conscious of seeing them.

It would be wrong to think, however, that either the amygdala or the insula was the 'seat' of consciousness, because with stimuli other than faces, completely different parts of the brain become active. If, instead of faces, people are shown cards with words written on them, something quite different happens. There is still a clear difference in the brain's response depending on whether the cards are seen for a long time or a very short time, but in this case two completely different parts of the brain are involved.[8] Again, it was found that if people were allowed to see the cards only briefly, they didn't think the cards had any words on them at all and they couldn't say what the word was. With long presentations, long enough for the subjects to report that they could see and read the words, their brains were very active on the left-hand side, in areas such as the left parietal cortex, already known to be associated with conscious reading. But when the cards were presented only briefly, at less than the critical value of 40 ms, then all this activity in the prefrontal and parietal areas was drastically reduced and became undetectable. In other words, particular parts of the brain appeared to be associated with the conscious perception of words, but these were different from that part associated with the conscious perception of faces.

A picture begins to emerge from these and other studies on brain activity in people performing tasks either consciously or unconsciously. Different parts of the brain are active depending on whether people report being conscious of something or not, but the part of the brain that becomes active varies depending on what they are doing. Reading words activates one part of the brain, whereas looking at human faces activates another part. There is no one place in the brain that is uniquely associated with being conscious. There is no equivalent of a chief executive sitting in an identified office switching consciousness on and off. There are no nuggets. The one indisputable conclusion from studies of brain imaging is that the connection between consciousness and brain in humans is far from straightforward. Consciousness can be associated with a generalized effect in many parts of the brain, as in anaesthesia, or it can be associated with activity in a particular area. But exactly which area will vary with different tasks and different situations because the brain has different areas for different specialized activities. Even a 'specific area for a specific task' view of consciousness is an oversimplification because for some states of awareness, several different parts of the brain need to be active at the same time.[9]

We are left looking at the pictures of brains from fMRI with a sense of wonder but an increased, rather than a decreased, sense of bafflement. We have not been able to pin down consciousness to particular kinds of activity, or particular areas of the brain, let alone explain where our subjective experience comes from. All we know is that a lump of greyish matter—about 1,330g in weight—houses everything we see and do and know. Out of this gelatinous mass comes consciousness. Somehow. But while we know how the individual nerve cells work and we are beginning to understand how they are linked up together, we have no idea how nerve cells, even billions of them acting together, could give rise to our consciousness. Even for us, the species we think we know from the inside, we have no idea how to get from objective studies of behaviour and what the brain is doing, to the subjective feelings we all experience. We still have no idea of how the sensations of smells and sounds are experienced and *felt*, or of why the pain *hurts* as well as consisting of activity in certain nerves.

All this is very bad news for understanding consciousness in other animals. If we cannot understand consciousness in ourselves, we are even further than we thought from understanding subjective states in other species. There is no identifiable organ of consciousness—no nugget—that we might look for in animals. Consciousness in us can apparently arise in many different ways, in different parts of the brain. We cannot even say we know what sort of neural activity is uniquely *correlated* with consciousness, let alone how these different sorts of activity all manage to *cause* consciousness.

This is in no way to belittle the importance of brain imaging as the most amazing technique for looking inside skulls and seeing what brains are up to. It has made it possible to make major progress in studying the living, active brain in a way that was completely impossible before the new technology came along. But this progress consists of advances in unravelling the 'easy' problems, such as how vision works,[10] whether the brain registers pleasantness of a taste in a separate place from where it identifies what that taste is,[11] or where in the brain decisions about what movement to make next are made.[12] But it has left the hard problem untouched.

All that brain imaging gives us is an exciting new way of measuring objectively observable behaviour, only in this case it's the behaviour of the brain. Only the one person actually in the brain scanner can know what it is like to experience, subjectively, what happens when a part of their brain lights up. Other people, the bystanders, can see the sudden change in the colours of the brain map but they cannot feel what it feels like. Brain imaging, for all the amazing detail it gives us of what brains do, does not allow us to study consciousness directly, any more than does measuring blood pressure, heart rate, adrenaline secretion, or even facial expressions. Consciousness itself remains elusive and subjective.

The plain and rather disappointing truth is that, even today, the only sure way that scientists have of studying consciousness in humans is to ask them questions such as 'Did you feel that?' Or 'Did you see that?' Note the extreme fragility of this approach. It relies completely on the assumption that when a person says they are conscious of something, they are conscious of it. There is no independent objective evidence that they are. The 'scientific' study of consciousness rests solely on the subjective report of the one person who is having the experiences. Hardly a model for the objective study of consciousness in non-humans.

Reports of consciousness are also notoriously dependent on people's moods and motivations. If people are uncertain about, say, whether they saw something, they might decide to be cautious and say they didn't, or say they had seen it because they thought that was what the experimenter wanted to hear. But the real problem is the uncheckable and therefore unscientific nature of subjective experiences themselves. What a person says about their subjective experiences is, in one sense, perfectly objective data. Their words are a form of behaviour that can be recorded objectively and heard by different people, in just the same way that their facial expressions can be. But what they *feel* and are consciously aware of when they say those words, that remains known only to them. You have only to think of people who may be unable to speak but may still be experiencing a great deal to realize that the distinction between what is said and what is experienced is very real indeed. Stroke victims or people who have lost the means of expressing themselves may still be conscious and experiencing a great deal. We do not know what people in a persistent vegetative state, who are 'locked in' to their bodies and unable to move, actually experience,[13] but it is possible they are conscious even though they cannot tell us. Here, the distinction between what is said and what is experienced consciously appears in a stark and terrifying form. We come face to face with the possibility of conscious experience inaccessible by words.

In our day-to-day lives, we do not let this distinction bother us unduly and for the most part, with normal healthy people, we are prepared to ignore the philosophical problems that consciousness presents and simply assume that other people do have conscious experiences much like our own. If I see you writhing in pain and crying out, I know that if I were behaving in that way, I would be subjectively feeling great pain so I make the assumption that you, too, are experiencing pain. I don't know this for certain but I ascribe my feelings to you by analogy because you are 'like me' in your anatomy, behaviour and, I am happy to assume, in what you feel.

So the scientific study of human consciousness is not quite as squeakily scientific as it might appear. Even with other humans, with the wealth of words that language has given us, and the dynamic brain images that modern technology provides, even then, when it comes down to it, the only way we have of inferring consciousness in other people is by asking them what they feel and then, by simple analogy, assuming that what they say they are experiencing is actually what they are experiencing. And if we do make that assumption and use their words as a reliable guide to what they are experiencing, what consciousness actually is remains as infuriatingly elusive as ever.

Unfortunately, by trying to creep up on animal consciousness by the backdoor of what we thought we definitely knew about our own consciousness, we find ourselves even further away from our goal than ever. The brain imaging technology that reveals the workings of the human brain in such an exciting way turns out to be exactly that and no more. It tells us about the objective working of the human brain and which areas are active at any one time. It does not make the study of consciousness any more objective than studying facial expressions or behaviour, or other indirect 'markers' of experience. Brain imaging can only tell us about consciousness by using a subtle form of 'cheating'—namely, asking people what they feel and using this as a substitute for studying their conscious experiences directly. And, as if this were not enough, brain imaging studies have told us the one thing we didn't want to hear—that conscious states are complex and cannot be tied down to particular areas of the human brain. There are no neural nuggets of consciousness that we can search for in other species. If we thought that human brain imaging would tell us what to look for in our quest for consciousness in other species, we have so far been turned away empty-handed.

The key question about animal consciousness (and there could be as many different forms of animal consciousness as there are species) is whether animals have subjective experiences that we would recognize as pain, pleasure like us, or fear. What we want to know is whether, inside an alien-looking skull, there is an 'I' that experiences something that we would recognize as consciousness. If there is, then we might try to build a kind of bridge and learn enough about other animals so that the gap becomes manageable and we can cross over with some confidence and understand their experiences from what we know of their biology. This is basically what we do with other individuals of our own species. But there is no point simply shouting 'Is there anybody there?' across the species gap and expecting some sort of answer, when we don't even know what sort of answer would count as showing that there was. So far, the study of human consciousness has not been a particularly helpful source of information. And as if this were not enough, recent studies have now dealt an even more serious blow to the search for consciousness in other species.

That blow has come from the increasing realization that there are many activities that we human beings do without being conscious of them at all. Just think of how much you do unconsciously, such as breathing or driving a car. A good driver has a foot on the brake pedal long before consciously realizing that a child has run out in front of the car. A pianist whose fingers fly over the keyboard in complex patterns does not do so by consciously remembering which notes come next, but leaves it to the extraordinary unconscious muscle memory system stored in the cerebellum. In many examples of skilled movements, an 'autopilot' takes over, leading to quicker responses and far more complex coordination than would be possible if every movement had to be consciously thought about.

Of course, when we are learning a skill we are all too conscious of our lack of ability as each step is slowly and painstakingly worked out.[14] At first, each note on the instrument or step in the dance routine requires intense concentration and conscious thought. Learning to drive a car seems initially to require thinking about more things at once than is humanly possible. But then we practise and, as we repeat the actions and become more experienced, our conscious minds can relax and hand over to our much more skilled unconscious system. Even I, a very mediocre pianist, have watched in astonishment as my left hand moves up and down the keyboard as I play a well-practised piece. 'I' have no knowledge of what the next chord is, but my left hand, almost like the hand of another person, lands steadily and surely in the right place at the right time.

Many human actions can be achieved by either a conscious route or an unconscious one.[15] This is true for driving a car, playing sport, or even just breathing. Most of the time, we breathe without consciously thinking about it. But if you were held under water and in danger of drowning, you would definitely start thinking about your breathing. Where your next breath was coming from would be uppermost in your conscious mind, whereas usually it is pretty low down. There is not just one 'breathing pathway' in your brain. There are different routes to the same behaviour (in this case breathing), some of which involve consciousness and others that do not.

A very striking demonstration of this is the phenomenon called 'blindsight', which is shown by people who have quite specific brain damage that affects their vision. Blindsight patients have perfectly normal eyes but they have damage in one of the main visual pathways leading from the eyes to a part of the brain that normally deals with visual information (known as V1, located in the visual cortex). This means that they can 'see' (with their eyes) but are 'blind' if they are looking at something normally processed in this part of the brain. Usually this only affects a particular part of their visual field so that they can see and report on objects perfectly normally as long as the objects are in one half of their visual field (the half that does not send information to the damaged bit of the brain), but they are 'blind' if the same object is placed in the other half of their visual field, when they cannot see anything at all. Or at least they say they cannot see anything. Larry Weiskrantz worked with a blindsight patient known as DB,[16] who became famous for what he could do visually even though he himself was certain that he could not see. For example, DB would be shown an object in his 'blind' field of view. He would reply that he couldn't see anything. However, if he were asked to guess what the object was, he was surprisingly good at identifying it, so he was not guessing at random. He seemed to have some knowledge of what the object was, even though he was not aware of being able to see it. In fact, he was very surprised at himself when he came up with the right answers, as he was quite sure he couldn't see anything at all.

DB and other blindsight patients can even perform actions that seem to need vision at the same time as swearing that they cannot see what they are doing. For example, they might be asked to post a letter through a slot. The slot could be either vertical or horizontal, so needing the letter to be pushed through at quite different angles. Despite saying that they can't see any slots at all, their hands turn the letter appropriately to going through a vertical or horizontal slot. Consciously, they cannot see what they are doing, but unconsciously, their hands are visually guided to do the right thing.

What is so striking about blindsight is that 'seeing' seems, on the face of it, to be the most conscious of processes. We see the light. We see what someone means. In so many cases, we use the word 'see' to mean being consciously aware or thinking about something. But here we have 'seeing' without conscious awareness. We have visually guided behaviour that does not seem to enter consciousness. In other words, the performance of a visual task (such as orienting a letter in the right way to go through a slot) does not have to involve conscious seeing, although in some circumstances it might.

Even more remarkably, humans can learn the rules of complex games without consciously knowing what those rules are. In a recent study of this,[17] groups of students agreed to take part in a project where they had to press one of two keys on a computer keyboard. Each student was told that if they pressed the correct key, they would receive £1 but if they pressed the wrong key they would have to pay a fine of 50p, but they were not told what made a key 'correct' or 'incorrect' on a given round of the game. They had to work that out for themselves. The only other thing they were told was that their computer keyboard was linked to those of other people in their group and that what these other people did might have an effect on what was correct or incorrect for them. Each student then had to make a series of 50–200 decisions about which key to press.

As they made more and more decisions, their behaviour gradually became more cooperative, in that towards the end of a series they were more likely to press a key that resulted in the whole group gaining than they were at the beginning. But the truly extraordinary thing was that they were not consciously aware of how they achieved this. They were unable to say what they thought the rules of the game were and they were not conscious of changing their behaviour in accordance with any rule they could think of. They certainly could not explain how their behaviour or that of other people was influencing how they decided which key was 'correct' for any given decision.

This ability of humans to perform even quite complex tasks without necessarily being conscious of what they are doing causes a real problem when we turn our attention to the possibility of consciousness in other species. If we can show that a behaviour can be conscious or unconscious when we do it, how do we know that another species, showing similar behaviour, is either conscious or unconscious? Other animals may be 'like us' in what they do, but like us when we are not conscious of what we are doing. The fact that we humans can do so much without being conscious causes an enormous problem to those who think that similarity of behaviour between humans and non-humans necessarily means similarity of conscious experience. Of course, other animals may be consciously aware in the same way that we are, but it is extraordinarily difficult to know one way or the other. The more we learn about the unconscious control of human behaviour, the more difficult it becomes. The difficulty stems from a very curious fact about the evolution of our brains. Brains seem to accumulate more and more different ways of doing the same thing.

From an evolutionary point of view, brains are very odd. Most parts of the body that have changed over evolutionary time and become adapted to a new function, have lost or compromised their previous function. For example, a horse's hoof, beautifully adapted to fast running over hard ground, no longer has fingers, even though the reptilian ancestors of horses, long ago, had five digits. Penguins, adapted to fast underwater swimming, still have wings but they have lost the ability to fly. Large flightless birds, blind cave fish, and legless lizards are all testimony to what appears to be an evolutionary rule: animals that specialize for one environment appear to lose some of the abilities that their less specialized ancestors had. The best swimmers cannot be the best fliers, the fastest runners cannot be the best diggers, and so on. Natural selection appears to be about compromise and gaining one capacity at the expense of another.

Almost uniquely, the brain appears to violate this rule. The history of the vertebrate brain is almost entirely a history of expansion of different parts and reorganization of what is already there.[18] Very little has been lost. Throughout evolution, from fish to human beings, the brain has simply added bits to itself without getting rid of anything old, rather like a person who adds solar heating to a house already heated by open wood fires, an electric immersion heater, and gas central heating, without throwing anything out.

What this means in practice is that we humans have all the evolutionarily old bits of the brain—the bits that fish and reptiles have—and we also have other more recent evolutionarily additions as well. The parts of our brain that control heart rate and breathing (brain stem and cerebellum) are still referred to as the 'reptilian brain' because they are essentially the same structures as are found in reptiles. We are still living with brain structures that our remote reptilian ancestors had hundreds of millions of years ago. They served them pretty well for meeting their basic needs and there was no selection pressure to get rid of them. We are also living with what is called the limbic brain (amygdala, hippocampus, etc.) that we share with all mammals and that is particularly concerned with emotions. The neocortex, the part of our brain associated with thinking and more complex activities, was simply added on top of these already existing structures. Extension and reorganization of what was already there seems to have been the way our brains have evolved. Rather than dispensing with structures that were working perfectly well, some parts of the brain were expanded and new connections between old structures were used to give the brain new abilities. We have extensive connections between the different parts of our brains, between the evolutionarily old brain stem and the relatively recent neocortical parts, between the emotional brain and the thinking neocortex, so that our brains function as coherent wholes, despite their many additions and reconstructions.

The extensive connections between the different parts of the brain and between individual nerve cells within a particular region are what give brains their extraordinary computing power. The human brain has 10,000 million nerve cells but each one may have thousands of connections to other nerve cells. This means that the activity of a particular part of the brain could be radically changed without the necessity for any major anatomical structure to evolve, but just by changing which neurons were connected to which other ones. For example, in the evolution of human language, it was not necessary to evolve a brand new structure in the brain. All the many areas of the cortex that are involved in language are present in primates that don't have language, such as chimpanzees. What has happened is that different existing parts of the brain have now connected up into an interdependent cortical network that (in ways we don't yet understand) gives us the ability to use and understand language.

This evolution of brains by accumulation and modification, rather than by loss and rebuilding, means that we humans still have many of the same brain structures as other animals. We have most in common with other mammals but we still have much in common with reptiles and even fish. The evolution of newer brain structures simply means that we humans have just got more ways of dealing with the same old problems of staying alive than other animals have. We have conscious ways and we have unconscious ways. But what we do not know is which brain structures are necessary for the conscious routes to be taken. This is because, as brains evolved, many changes of function (including, for all we know, changes from unconscious to conscious processing) took place through subtle reorganization of what was there already. By making new connections between existing structures, animals evolved new abilities. The old brain areas were able to perform new functions simply by being connected up in new ways.

So, when it comes to trying to understand other species, the recent advances in human consciousness leave us more bewildered than ever. Are animals conscious like us because they have many brain structures in common with us, or are they not like us because they lack some crucial pathway that prevents them from going that extra mile into conscious experience? Are they like us but only when we are using an evolutionary old unconscious route, or does a spark of what we would recognize as conscious experience flare in even the oldest of brains? On the basis of what we now know about human consciousness, both of these completely opposite views can be, and indeed have been, put forward as serious hypotheses.[20] The elusive source of our own consciousness and its infuriating refusal to be tied down to particular neural structures leaves us, at the moment, completely unable to distinguish between these quite opposite views of animal consciousness. The more we have tried to find out about human consciousness, the more difficult animal consciousness has become. The hard problem just got a whole lot harder.

6

EMOTIONAL TURMOIL

Emotions can seem to us the most conscious of all conscious experiences. We quake with fear or we are racked with guilt. We are consumed with joy or overwhelmed with sadness. So if ever there were a point at which we might begin to understand the conscious experiences of other species, that point might well come through the experience of basic raw emotion. Perhaps we should forget all the clever things that animals can do and pay more attention to the ability to feel emotions. You don't have to be particularly clever to experience pangs of hunger or to be afraid of the dark. Rationality and thinking might be 'explained away' as occurring without consciousness, but can emotions be so easily dismissed by explanations that operate without the light of consciousness?

Emotions also have a particular importance for animal welfare. The belief that animals can both enjoy pleasurable emotions and suffer from unpleasant ones such as fear, grief, boredom, or frustration is what gives many people their chief reason for being concerned about animal welfare in the first place and is what gives animals the moral edge over plants or valuable works of art. The unpleasant emotions we call suffering become the foundation for moral decisions and for deciding that it is right or wrong to treat other humans or animals in particular ways. 'A full-grown horse or dog, is beyond comparison a more rational, as well as a more conversable animal, than an infant of a day or a week or even a month, old', wrote the philosopher Jeremy Bentham in famous words that echo down the centuries, 'But suppose the case were otherwise, what would it avail? The question is not, Can they *reason?* nor, Can they *talk?* but, Can they *suffer?*'[1]

The study of animal emotions has a distinguished and honourable champion in the form of Charles Darwin, whose book *The Expression of the Emotions in Man and Animals* stressed the evolutionary continuity between us and other species.

Darwin wrote in a quite uninhibited way about angry chameleons, disappointed chimpanzees, enraged lions, cowardly gamecocks, irritated snakes, and 'appallingly furious' moose-deer. He described dogs as being, at different times, cheerful, terrified, or in a state of perplexed discomfort. Even insects, he wrote, 'express anger, terror, jealousy and love by their stridulation'.[2]

But although Darwin emphasized over and over again how like other species we are, he was far too canny to assume that every time we or another animal 'expressed an emotion', either we or they are necessarily conscious of what we are doing. One of the main concerns of his book was to distinguish which expressions of the emotions developed through will and consciousness and which did not. His conclusion[2] was that 'The far greater number of the movements of expression, and all the more important ones, are, as we have seen, innate or inherited; and such cannot be said to depend on the will of the individual.' He did believe, however, that for some expressions, such as the human habit of drawing down the corners of the mouth when distressed, there was an element of learning so that consciousness and will were initially involved, even if later in life this became a habit. He added: 'not that we are conscious in these or in any other such cases what muscles are brought into action, any more than when we perform the most ordinary voluntary movements.' Darwin was thus fully aware of the distinction between the behavioural expression of an emotion and the conscious awareness that might or might not be behind it. At the time of expressing an emotion, the control of movement could be quite unconscious, even in us.

The distinction that Darwin made between the external expression of an emotion and the inner experience that may or may not accompany it is still widely recognized today in discussions of both human and animal emotions, although not everyone sticks to the distinction quite as clearly as he did. Those who study human emotions see them as having three separate components.[3] The first component includes all the bodily changes that happen when we feel an emotion very strongly, such as our heart rate going up and hormonal changes occurring, or our face flushing pink. These we humans have in common with other species and they can be observed and measured quite objectively.[4, 5] The second component includes all the behaviour and facial expressions a person might show when they are in the grip of an emotion. These, too, we share with other species, as Darwin laid out in such illuminating detail. The third and crucial component is the conscious experience of emotion that we humans know we have—what is going on subjectively while the outward and visible components of emotion (the physiology and the behaviour) are happening.

The question of whether this third component is also shared with other species is just as difficult for us to decide now as it was in Darwin's day and raises just the same difficulties as anything we discussed in Chapter 5.

Given the central importance to animal welfare of the idea that animals actually do experience emotions, it is widely advocated that these difficulties should be ignored and we should simply assume that animals do have this third, conscious, component of emotion even though it is impossible to be sure.[6] Michel Cabanac has argued that since many animals show a whole gamut of the same physiological symptoms that we show when we feel an emotion such as a rise in temperature, blushing, the hairs rising on the back of the neck, and a racing heart, the most plausible interpretation is that they too subjectively feel emotions.[7] He argues that they are so like us in their physiology that it is utterly implausible to argue that they are not like us in what they experience.

Cabanac believes so strongly that many animals do have all three components of emotion that he is prepared to draw a line in the animal kingdom between animals that do and those that don't feel emotions. He draws his line between amphibians and reptiles on the grounds that handling a lizard causes its heart rate to go up and its body temperature to rise whereas neither of these physiological symptoms of emotion occur in either fish or toads. His conclusion is that lizards are conscious but that fish and toads are not. The subjective experience of emotion, for Cabanac, evolved because it allowed reptiles, birds, and mammals to make much more sophisticated choices between the various options open to them, such as whether to feed, seek warmth, or run to cover. He believes these animals make such choices by experiencing positive and negative emotions (which he refers to as pleasure and displeasure) and then choosing the option that gives them the greatest pleasure.

Cabanac gives a number of examples of the striking similarities between the ways in which we respond to situations we consciously find pleasurable and the way animals respond to the same situations. His implication is that these examples give weight to the view that animals have similar conscious experiences. One of his experiments involved giving rats the opportunity to press a lever that altered the temperature of a small disc placed on their skin.[8] The rats learnt to press the lever with no reward except to be able to change the temperature of the disc, but they only did so if the existing temperature was either several degrees higher or several degrees lower than their own natural body temperature. However, if the existing temperature was much closer to their own body temperature, they pressed the lever less vigorously and if it were the same as their own body temperature, they did not bother to press it at all. The rats' behaviour with their lever was strikingly similar to the behaviour of people dipping their hands into a bath of water and being asked how pleasant or unpleasant they found the temperature of the water. People in an uncomfortably hot room reported that cool water was very pleasant to the touch but water of the same temperature in a room closer to normal human body temperature was described as only moderately pleasant. In a cold room, it was the warm water that they said was more pleasurable, with the cool water now being described as unpleasant. The verbal reports of what people were feeling (very pleasant, moderately pleasant, neutral, moderately unpleasant, very unpleasant) so exactly paralleled the behaviour of the rats (lots of pressing on the lever—less—none—less, and lots again) that Cabanac argues it is but a short step to saying that the rats are, by their lever pressing, effectively telling us that they too experience some temperatures as pleasant and some as unpleasant. Add to that the similarities of emotional responses, such as increase in pulse rate and body temperature, and he concludes that there is a convincing case for subjective emotional experiences. Cabanac has even trained a parrot to shout 'Bon!' (Good!) whenever it is given something it likes.[9] The fact that the parrot can use this word to describe new objects and new situations it likes he sees as evidence that birds, like us, have a generalized view of what pleasure is.

Cabanac may well be right in making his leap of analogy between ourselves and other animals in this way, but it is still important to recognize that the evidence on which he does so is just that—a leap, not scientific evidence. He has chosen to hop over the hard problem rather than try to solve it. He is not alone in being prepared to do so. Jaak Panksepp and Joseph LeDoux also both argue that humans share certain core emotions such as fear, rage, lust, panic, and play with other mammals, and that because these are associated with brain structures that we share with them, they feel these emotions just as strongly as we do.[10] The parts of the brain that are particularly implicated in emotion, such as the amygdala, are the evolutionary older parts of the brain that we share with other mammals. Panksepp calls them the 'primary process ancestral birthrights of all mammals'.[10] Our neocortex, the part of the brain we proudly say is responsible for specifically human abilities such as language, is simply stuck on top of these ancient emotional circuits. The older, basal parts of the brain associated with basic emotions such as thirst, hunger, and fear, we share with an even wider range of animals, such as reptiles, and their function is very similar between us and them.[11, 12] The implication is that we share emotional experiences with them too.

Linking subjective experience to brain structure, behaviour, or brain chemistry in animals would be relatively easy if the three components of human emotion always went together. If every time you (subjectively) felt an emotion, someone else could reliably pick up correlated changes in your behaviour and the physiological changes going on inside your body, we could begin to use these observable changes as surrogate indicators of conscious experiences. We could look for Hormonal Correlates of Consciousness (HCCs) or Behavioural Correlates of Consciousness (BCCs) in the same way that neuroscientists have looked for the Neural Correlates of Consciousness. We could then see if other species showed these indicators and infer, by a not too great a leap of analogy, that they had conscious emotional experiences too.

But emotional consciousness has refused to be tied down to particular parts of the brain or to hormones, heart rate, facial expressions, or any specific behaviours, just as steadfastly as any other kind of consciousness then NCCs. As we saw in Chapter 5, there are no consciousness 'nuggets' in the brain, no neural activity correlating perfectly with what people experience. HCCs and BCCs have proved, if anything, even more unreliable as indicators of consciousness than NCCs. The three components of emotion—the ones we can observe (behaviour and physiology) and the one we cannot observe but really want to know about (the conscious experience of emotion)—just simply don't correlate very well in humans, leaving us just as uncertain as ever what to rely on when we look at other species.

For example, most of the bodily changes that occur when we subjectively feel an emotion—such as in skin conductance, heart rate, facial temperature—are very similar whether the emotion we are feeling is anger, fear, or even happiness.[13] If all you knew about another person's emotional state was from these bodily changes, you would be very hard put to say whether they were feeling angry, afraid, or just very, very happy.

People riding on roller-coasters provide a good example of how the outward and visible signs of emotions are a very unreliable guide to what they are actually feeling. People scream as the roller-coaster descends a curve in the track, their knuckles white as they grip the seat in front of them. Their hearts are racing, the adrenalin courses through their veins, and their faces are contorted. But someone standing on the ground watching them, or even monitoring their bodily changes remotely, would be hard put to it to say whether the emotion they were experiencing was fear, excitement, or sheer pleasure. Indeed, someone could move from pleasure to genuine fear with very little, if any, change in behaviour. The scream of exhilaration could turn to a scream of genuine fear and the bystanders could be completely unaware of it.

The reason why subjectively felt emotions are not conveniently tied to particular external signs of behaviour or internal bodily changes is that emotions themselves are not arbitrary. It does not just so happen that when we get angry we have a lot of anger symptoms and when we feel 'happy' we have a different set of happiness symptoms. Emotions prepare us for actions and many of the actions we perform make the same demands on the body and so need the same sort of preparation. Getting angry with someone means that we might have to fight them or run away if they start threatening us. Fighting and flight both require the body to be in a state of alert, to have plenty of oxygen available for moving the limbs (greater blood flow achieved through a higher heart rate), and fuel available in the form of glucose to power them (hormones releasing this into the blood). No wonder anger and fear are difficult to distinguish physiologically. Running races, playing sport, or simply leaping around for joy require many of the same preparations for activity as fleeing for your life, so that it is not surprising to find that these same preparations are common to many different emotions and are therefore unreliable guides as to what subjective experiences might be accompanying any one of them. And the fact that they are such unreliable and inconstant companions to the conscious experience of emotions in ourselves means that they cannot be relied upon to tell us about the emotional experiences of other species, beyond the fact that their bodies, too, prepare them for action.

But perhaps it is not that surprising that hormones and pulse rates, sweaty palms and blushing faces are not the unique hallmarks of any particular emotional experience. Conscious experiences are a product of our brains, not our hearts or our livers or our stomachs, although 'funny feelings' in these organs may affect what sort of experiences our brains come up with.[14] It is only poetic licence that leads us to speak as though emotions come from the heart, not the head, and this is left over from a time before people understood the central role of the brain in all our experiences. The search for the conscious component of emotions must, therefore, be centred on our brains. But here we find that the situation is so complex that, once again, studying human consciousness is a setback rather than a help in deciding whether other species have the experience component of emotion too. Unfortunately, the hard problem has not gone away just because we have switched from intellect to emotions. On the contrary, we now have disconcerting evidence that humans can have emotions without consciously experiencing them.

As we saw in Chapter 5, our unconscious minds can control a whole range of actions from well-practised playing of a musical instrument and driving a car to routine breathing, without our being consciously aware of what we are doing. However, the abilities of our unconscious minds do not stop at sport, music, and breathing. They also extend to having emotions without our being aware of them.

If people are shown a photograph of either a happy or a sad human face for such a short period of time (less than 40 thousandths of a second (ms)) that they are not conscious of having seen a face at all (at least by what they report), nevertheless the expression on the face has an emotional effect on them.[15] A happy face makes them more likely to say that a previously unfamiliar and neutral symbol, such as a Japanese word, means something happy or pleasant even though they have no idea what it really means. But a sad face makes them interpret an unfamiliar symbol as something depressing or unpleasant, even though in neither case are they conscious of having seen a face at all. In other words, the emotional interpretation people give to something can be influenced by something they are not even conscious of having seen.

People with 'blindsight' show an even more dramatic separation between the behavioural expression of an emotion and the conscious experience of it. 'Blindsight', as you will remember from Chapter 5, is an affliction in which patients have brain damage that prevents them from seeing in some parts of their visual field even though they have perfectly normal eyes. It is their brains that make them blind, not their eyes. So if a picture of a person with an angry, fearful, or happy face is presented to them in their 'blind' field they will say they cannot see any faces, and so have no idea what expression the face they cannot see might have. But the amazing thing is that at the very time when they are denying being able to see any face at all, they are mimicking the expression of the face in the photograph—pulling an angry face if the picture is of an angry face, and smiling if it is a happy one, and so on.[16] They report that they cannot 'see' the face consciously but unconsciously they must be seeing it because they are making the same facial expression, without being conscious of what has brought about their change in emotional behaviour.[17]

This role of the unconscious in the expression of human emotions makes the study of emotions in other animals much, much more difficult than we might have hoped. It leaves us with the distinct possibility that other animals might be like us in having emotions, but only like us when we have emotions that are unconsciously, rather than consciously, experienced. Unconscious emotions in ourselves leave us with a raft of yet more unanswered questions about how we can decide when and if animals have conscious experiences for exactly the same, infuriatingly persistent reason we have discussed before—we do not know how conscious, rather than unconscious, emotions arise in ourselves and therefore we do not know what to look for in other species as 'evidence' for consciousness in animals. Consciousness remains the enigmatic passenger on the bus. We cannot tie it down to particular brain activity or hormones or facial expressions although it is often (but not always) correlated with these observable events. We do not understand the connection between brains and emotional consciousness any more than we understand the connection between brains and being conscious when we think rationally. Gazing perplexedly at the animal kingdom, we could as logically conclude that almost all animals have consciously experienced emotions as that none of them do. Darwin's enraged lions and jealous insects might be consciously experiencing anger and jealousy, but, on the other hand, they might just be showing the behavioural and physiological symptoms without experiencing anything, just as we do at times. Darwin himself admitted he had no idea.[18]

Even more frustratingly, every single function that has been proposed for consciously experienced emotions could equally well be carried out without consciousness. For example, one proposed function of emotions is that they motivate us to do things and keep us going when the original stimulus has disappeared.[19] We don't suddenly stop feeling the emotion of fear just because we can no longer see an enemy. The sight of a cheetah may cause the initial alarm in an antelope and cause it to run away. But as the antelope turns, it can no longer see the predator behind it. It would be a notably unsuccessful antelope that stopped running just because it could no longer see the cheetah. Having a mechanism that not only triggers escape behaviour but then keeps the escape behaviour going until the danger is really over is very important to survival. Emotions, characterized by the fact that they outlast transient stimuli, fulfil this function very effectively. But there is nothing to say that the emotion of fear has to be consciously experienced. An increase in heart rate can occur very quickly at the sight of a predator, putting the animal on alert and triggering escape. Hormones are then ideally suited to maintaining the state of alert. They are secreted into the blood, circulate to the organs, and have their effects over a longer timescale. The combination of immediate response and slow hormonal switch off (which we can choose to label as the emotion of fear) is a beautifully adapted mechanism to get the animal out of danger and keep it there. The animal does not have to consciously be aware of anything. The fear, in other words, might or might not be accompanied by a subjective experience. The emotion works either way.

A similar 'take it or leave it' account of conscious experience can be given for another proposed function of emotion, namely that it allows different stimuli to be evaluated and compared and for choices to be made.[20] The choice might be between different foods or it might be between different courses of action such as whether to attack a rival or flee. Emotions provide a common currency for deciding which option to go for. Each option is assigned a value on a scale that has very pleasurable at one end and nasty or punishing at the other, and the options are compared even though they may be very different. A sweet fruit might be preferred over a sour one and so given a higher value. But if the sweet fruit could only be obtained by climbing a high tree, some value would have to be put on the unpleasantness and danger of having to climb the tree. How to decide between a sweet fruit only obtainable by climbing and a less sweet one that was more easily available? Somehow climbing trees and sweetness of fruits have to be compared even though they are very different. By assigning an emotional value to each, emotions provide a common currency for evaluating different options and deciding which is the most valuable at any one time.

A real-life example can be seen in ponds every spring. Male newts display to females underwater. Successful courtship from the male newt's point of view is a female newt picking up a packet of his sperm and using it to fertilize her eggs. But persuading a female newt to do this requires a considerable period of vigorous tail-waving by the male first. Tail-waving uses up oxygen and, as newts breathe air, an enthusiastic male newt can soon find himself out of breath and faced with a serious dilemma. Sex or breathing. He has to decide. He has a delicately balanced decision-making mechanism that initially decides in favour of sex, inhibits breathing, and keeps him underwater far longer than usual.[21] But there are limits and if the female takes too long to respond, he decides the other way and dashes to the surface for air, possibly losing the female altogether. Here we see a finely tuned decision-making mechanism, adapted to deal with a conflict between two opposing demands, but not necessarily tied to a conscious experience. The male newt might be consciously deciding between increasing his number of offspring and his own short-term survival. But then again, his decisions about which of his emotional states are strongest at any one moment could be quite unconscious. The functions of choice and decision-making could be carried out either with the full pano-ply of conscious experience or completely without it.

When an organism makes a choice, this in itself says nothing at all about whether there is a conscious mind making the choice. A plant 'chooses' to grow towards light rather than away from it and may even 'want' to grow upwards so much that it pushes paving stones out of the way in the process. But this can all be done without consciousness. We don't contemplate the structural damage done to a building by the roots of a tree and conclude that the tree consciously wanted to do anything at all, despite the 'effort' it obviously made, any more than we say that water 'wanted' to burst our pipes because of the huge forces it had to exert to do so.

Similarly, when we come across a mechanical or electronic system that closely mimics the behaviour of pleasure-seeking rats and humans, we don't assume that it must have conscious emo-tional experiences just because it seems to have goals. The central heating system of a house that automatically switches itself on when the temperature drops below a set level has a goal of what the temperature is 'supposed' to be, but it does not achieve that goal consciously. The heating system's goal is a particular tempera-ture inside the house. The further the temperature of the house is away from this goal, the harder the boiler will work. When the goal is achieved the boiler switches itself off. We could say it was no longer motivated to work. It would be trivial to program a 'speak-ing house' that told you it was very unpleasantly cold when there was a big discrepancy between the set point temperature and the actual temperature, and to report that everything was very pleas-ant when the actual internal house temperature matched the desired set point. The fact that there is an analogy between what we do when we are cold (take steps to get warm, switch on heaters, etc.) and what a centrally heated house does when it is cold (switches on heat) does not justify the conclusion that the house, even though it might be speaking to us in English, feels anything at all.

Sorting out whether animals consciously experience emotions is made all the harder by the emotional impact of the word 'emotion' itself. It is an extreme example of the seductive power of words we discussed in Chapter 2. It is so laden with the implication that to have an emotion must include the conscious experience of emotion that it is sometimes hard to separate out the three components or to believe that there could really be any such thing as an 'unconscious emotion'. Even when the evidence is spelled out, as with people showing emotional behaviour without conscious experiences, it is all too easy to slip back into thinking that just by saying we are studying 'emotions' in animals means that we have already established that they have conscious experiences. The trouble with the word 'emotion' is that it can be used to describe both the behavioural and physiological symptoms that are clearly in the realm of science and also the subjective experiences that might or might not be going along with them. People can't usually be bothered to say 'Emotion, by which I mean the physiological and behavioural only' or 'Emotion meaning all three components'. So two people can both be talking about emotions and can mean completely different things. Or even mean the same thing but on different occasions. Or even not be clear themselves in which sense they are using the word.

For that is the real danger of the word 'emotion'. Because it is both a scientific term and a word in everyday use, it can confuse the way we think and make us believe that the hard problem has either been solved or is just an insignificant hop away from being solved. It therefore encourages anthropomorphic thinking of the very worst kind—inadvertent anthropomorphism. People use it and, even if they are clear in their own minds what they mean, they inadvertently mislead their audience into thinking that we already know that animals feel and experience things just as we do when we are being emotional. The word 'emotion' itself encourages uncritical anthropomorphism.

The same is true of a lot of other words associated with the study of emotion—words like 'mood' and 'positive affective state' (scientists' way of describing pleasure!). 'Mood' is like emotion but usually taken to mean something even longer lasting. A change of mood can continue over hours, or even days, after the original stimulus has disappeared. For example, if an animal lives in an environment where there are a lot of predators, its behaviour may become adjusted to constant danger. It might be particularly flighty or be reluctant to leave the shelter of hedges. Its 'mood' is antipredator. But if there were no predators around for some time, its mood might change and it might become 'bolder', not taking flight so easily and taking greater risks. A particularly adaptive change of 'mood' has been described by Nick Davies in his beautiful studies on the behaviour of small birds towards cuckoos.[22] Cuckoos are nest parasites—that is, they lay their eggs in the nests of other species, such as reed warblers. Having a cuckoo egg in the nest is a disaster for the reed warbler parents because the cuckoo egg hatches out before their own eggs and the cuckoo chick then ejects all the reed warbler's own eggs. A cuckoo successfully laying an egg in a reed warbler's nest potentially means that the reed warbler parents will have no chicks of their own that year. To avoid this reproductive disaster, reed warblers have evolved a variety of anti-cuckoo defences, including being very discriminating between their own eggs and those of cuckoos. If they detect a cuckoo egg in their nest, they throw it out. Over time, there has been an elaborate arms race between cuckoo and host, with the cuckoo eggs evolving to look more and more like genuine reed warbler eggs and the reed warblers becoming better and better at telling the difference.

But there are problems with being too discriminating. If reed warbler parents err on the side of being too discriminating, and the female just happens to lay a slightly unusual-looking egg herself, then that genuine egg might get thrown out. Remarkably, reed warblers have found a way of adjusting how discriminating they are depending on whether or not they have seen a cuckoo near their nest, thus making it more likely that there will be a cuckoo egg around. The sight of the cuckoo not only arouses them to attack when the cuckoo is present, but their anti-cuckoo mood lasts for several hours after the cuckoo has disappeared. For the rest of that day they become more likely to eject 'odd' eggs from their nest. Describing this as an anti-cuckoo 'mood', characterized by a shift in how eggs are responded to, is a neutral way of describing what is going on and makes no implications about what the reed warblers are thinking or feeling.

But if moods are described as 'optimistic' or 'pessimistic' then it becomes more difficult to keep such implications at bay. Many animals show a shift in their choice between stimuli depending on what sort of environment they have been in prior to being tested, which seems to reflect a change in mood.[23] For example, starlings can be trained to associate different kinds of pots with either tasty or nasty food. If a starling has learnt that a pot with a black lid contains a preferred food item such as a mealworm and a pot with a white lid has a mealworm laced with quinine (which starlings reject) it can then be presented with a pot with a grey lid to see whether it treats grey-lidded pots as more like pots with black (nice) lids or white (nasty) lids. If a starling has previously been living in an enriched cage, it is more likely to respond to an intermediate grey-lidded pot as if it were like a black one and contained a tasty morsel. But if it has been living in a bare cage without enrichments, it is more likely to take the 'pessimistic' view and treat a grey pot as more like a white one and therefore unlikely to contain anything worth eating.[24] The long-term emotional state or 'mood' induced by the environment in which starlings have been living previously carries over and affects their choices in the experiment, even when they are no longer in that environment.

This is an important finding about the lasting effects that environments can have on the behaviour and emotional state of animals. But it says no more about whether those long-lasting emotional states are accompanied by conscious experiences than any other behavioural or physiological measures of emotion do. Even if animals are described as 'pessimistic' or 'optimistic' because their behaviour is affected over the long term by previous experience, this is no more an indication of what they are consciously experiencing than is the male newt switching his behaviour or the switch of mode on a computer or mobile (cell) phone.

So, it is time to grasp the nettle, face up to consciousness, and say that until we have answered the hard problem, consciousness cannot be tied to hormones or pulse rates or the ability to make choices or even particular parts of the brain. We therefore cannot say which animals are conscious like us. We can begin to answer some of the 'easy' problems (difficult though they are) such as how animals make decisions, how they choose between different options and how and what they learn. But we cannot yet answer the questions of which of these abilities bring conscious experiences along with them. All of them could be done without consciousness and many of them are done without consciousness even by us. And we cannot yet answer the really hard question of how or why on earth mechanisms that could apparently operate 'in the dark' ever gave rise to the light of consciousness. We can't answer it in ourselves and we certainly can't for other species.

Many people have argued that we should respond to the hard problem either by making out that it isn't really that much of a problem at all[25] or by criticizing anyone who insists it is a problem as dangerously pedantic with a mind still in the grip of an outdated behaviourist 'taboo'.[26] These arguments are understandable because of the central place that animal consciousness occupies in so many people's reasons for being concerned about animal welfare in the first place. Consciousness is for them the heart, the main reason for being concerned about animal welfare. If we care about animals, they continue, we should give animals the benefit of the doubt and assume they are conscious. They believe that the 'indirect' scientific evidence that many animals are conscious is so overwhelming that the burden of proof should no longer be on those who want to claim that they are, but on those who want to claim that they aren't. If you take this line, then questioning what we really know about animal consciousness can therefore seem like a direct attack on the very foundations of animal welfare itself.

The argument of this book, on the other hand, is that there is no proof either way about animal consciousness and that it does not serve animals well to claim that there is.[27] Both the argument that many animals do have conscious experiences and the argument that most of them don't come up against exactly the same implacable hard problem of consciousness. This is that we do not understand what consciousness is or how the activity of brains, even our own, gives rise to it. To turn our backs on such fundamental issues and say that we just need to give animals the 'benefit of the doubt' may seem like a pro-animals thing to do. It may seem like the best way to give ethical support to animals, but the advantage may be short-lived. It gives the impression that the best case for animal welfare rests on solving problems that so far have defeated the best and ablest thinkers down the ages. It enables people who do not care about animals to argue that the case for animal welfare is weak because it rests on untestable and unscientific claims about the nature of animal consciousness. It makes it look as though animal welfare 'science' is not really scientific at all.

The best and most convincing argument here is that we have no idea which animals are conscious because we have not yet solved the hard problem sufficiently well to be able to rule any animal either in or out of the consciousness club. The very best weapon against those who claim that only humans are conscious is to point out that this claim is invalid until the hard problem has been solved, which it certainly has not been at the moment. This is a small price to pay to keep the idea of animal consciousness alive and well for a whole variety of species. The hard problem deals with all theories of consciousness with equal ruthlessness. The consciousness club (or, more accurately, the potential consciousness club) has many more members if we acknowledge that we simply do not know when during evolution conscious experience kicked in than if we use a particular, not very convincing theory of consciousness to decide. Just as we cannot demonstrate that any animals other than humans are conscious, so we cannot rule them out either.

An even more powerful reason for facing up to the hard problem is that this provides the best possible defence against the increasing number of 'killjoy' explanations that are being put forward to explain examples of animal behaviour that have previously been claimed to be evidence of animal thinking. But, as Chapter 4 made clear, even killjoy explanations do not kill off consciousness. They only appear to do so if you believe that we understand consciousness sufficiently to say that we know how to recognize it or what gives rise to it; in other words, if you believe the hard problem has already been solved.

If, on the other hand, we acknowledge that we do not know whether consciousness is a property of the stupid or the clever, the emotional or the unemotional, the innate or the learnt, the immediate sensation or the foresightful plan, the language users or the grunters—if we acknowledge that we really don't know, then the possibility of consciousness in all sorts of species remains intact.

So, if you want to support animal consciousness in the face of attacks by killjoys, acknowledge the hardness of the hard problem. Be a pedant. Point out that killjoy explanations are to be welcomed because they clarify our ideas about what animals are actually doing, but say nothing about consciousness itself because the evidence that is being 'killed' isn't evidence for consciousness anyway. For all we know, consciousness is alive and well inside millions of different sorts of skulls, and even within brains that don't have skulls and that we would hardly recognize as being brains at all. We might even find consciousness in crabs or prawns or insects.[28]

But, above all, the most important reason for embracing the hardness of the hard problem is that it is an important step in establishing animal welfare as a scientific discipline. If animal welfare scientists get a reputation for claiming that they have evidence for consciousness when in fact they have no such thing, then the science as a whole will not be taken seriously. Animal welfare will be seen as a 'soft' subject carried out by people who do not understand what the difficult problems really are.

So, if animal welfare is to have a real impact on a world increasingly concerned with how to provide for human welfare, it must have a voice that states its case with a backup of solid evidence. Simply appealing to anthropomorphism is not enough. It may be enough for those who are already convinced that animals are like us in what they feel and experience. But it is clearly not enough for the large numbers of people who are so far unconvinced and whose main concerns are the welfare of human beings. To appeal to the many people who put human welfare as their top priority, way above anything to do with non-humans, by using unconvincing evidence that animals 'most probably' have conscious experiences, is not the best way to get them to take animals more seriously.

For animal welfare to be given higher priority than it currently enjoys in discussions of the future of the planet, it should not rely on animal consciousness as its only, or even its main argument.[29] If it tries to, it will be shot down for lack of evidence and animal welfare will suffer. Even though its proponents may think the evidence is there, its opponents do not and they are in the majority across the world. Animal welfare needs new and more powerful arguments to support it and it needs those arguments to be based on scientific evidence. This may be a disappointment to people whose ethics are firmly rooted in the belief that animals consciously experience pain, pleasure, and suffering, but there is nothing to stop you continuing to have these beliefs and continuing to use them as the bedrock of why you believe animal welfare to be important. The new arguments we are about to discuss can simply be added to the foundation you already have. But the great thing about these other arguments is that they can also stand alone. They do not require the hard problem to be solved or animal consciousness to be in any sense proved before they are valid. They appeal to people's self-interests, which gives them just the power they need to make the improved case for animal welfare. They make it much more likely that the voice of non-human animals will actually be heard.

7

ANIMAL WELFARE WITHOUT CONSCIOUSNESS

There is a wonderful feeling produced by finally surrendering to the hardness of the hard problem. The burden of having to 'solve' the problem of consciousness before being able to think about animal welfare is instantly lifted. It is no longer necessary to believe you have to find answers to problems that have defeated philosophers and scientists down the ages. Questions about the connection between minds and bodies or about where consciousness comes from can be acknowledged for what they are—difficult nuts to crack and ones that are beyond current science to explain. The explanatory gap is still there but it doesn't have to get in the way of thinking about the ethical treatment of animals. It is a great relief not to have to pursue impossible goals and to be able to rely instead on some much more down to earth and practical reasons for believing in the importance of animal welfare.[1]

But if we are not going to use conscious awareness of positive and negative emotions for the definition of good and bad welfare, we are going to need something to put in its place. If demonstrating that animals consciously experience grief, fear, pain, boredom, and the other emotional hells we refer to as 'suffering' is no longer going to be the only or even the main definition of 'poor welfare', then what are we going to use? If these emotions are not up to the job of making a good enough case for animal welfare, then what might be?

Although many, many words have been written about the exact meaning of animal welfare, and even more about how to measure it,[2] what most definitions have in common is that 'good welfare' has to include the animal being in good health. Animals that are riddled with disease, injured, dehydrated, starving, or exhausted are not in a state of good welfare by almost anyone's definition. We may want to add some extra animal requirements for the good life (Chapter 8) but on any definition, good physical health is the foundation of all good animal welfare. The corner stone. The starting point. Fortunately it is also the basis of the most powerful argument to place in front of people for whom animals matter least in comparison with human beings: whatever your view of animals, their health directly affects your health.

According to the World Health Organization, about 75 per cent of the new diseases that have affected humans over the past ten years have originated from animals or animal products.[3] Many have the potential to become global pandemics and many have already killed millions of people, such as the flu pandemic of 1918. Many more serious diseases such as anthrax and rabies are carried by animals, as are some of the most virulent food-borne pathogens such as *Campylobacter, Escherichia coli, Salmonella, Shigella,* and *Trichinella.* Animal health directly affects human health because so many animals have bodies like us and immune systems like ours and so can harbour diseases that can also infect us. The disease organisms can (and do) jump species barriers, find a new host that is not yet immune to them, and they can spread with devastating effect. The names of some of these diseases—bird flu, swine flu—carry their origin with them wherever they go. Everyone on the planet has an interest in making sure that they don't get going and that they don't spread, so everyone on the planet has a reason to be concerned about animal health, whatever their views about animal consciousness and however they think animal welfare should be defined.

A second, related way in which animal health directly affects human health is through food. The health of both food plants and food animals directly affects whether there will be enough to feed the 9 billion people that are expected to be living in 2050.[4] It also affects whether that food is of the quality that people want to eat, and the healthiness of their food is something people are prepared to pay for.

Tell people that meat is produced in a way that isn't very good for animal welfare and they may or may not buy it, if the price is right. But tell them that the meat is contaminated and might just possibly affect their health and they are very much less likely to buy it. A threat to their own health or to that of their children is a much more powerful reason for changing people's buying or eating habits than appeals to what is good or bad for animals. In a 2009 survey by the UK's Food Standards Agency (FSA) about what influenced people in the food they bought, about 20 per cent mentioned animal welfare as a factor, but over 60 per cent gave 'food that is healthy' as a reason for buying, and 40 per cent specifically mentioned food hygiene or a low risk of food poisoning.[5] (Interestingly, in the 2010 FSA survey of consumer attitudes, animal welfare moved up to joint fourth place on the list of people's concerns when food shopping, behind the price of food (a concern for 54 per cent of respondents), the amount of salt in the food (45 per cent), and the amount of wastage (42 per cent).) Animal welfare shares a 40 per cent mention with another human health issue, the amount of fat in the food. Americans tend to be even more health conscious in what they eat than Europeans[6] and are particularly concerned with how food relates to health issues such as obesity, antibiotic resistance, and diet-related diabetes.[7]

Even when people buy organic or free-range food, this is often because they perceive it as healthier for them to eat rather than just because they believe it is better for animal welfare. A Nielsen survey of 27,000 consumers in Europe, Asia Pacific, Pakistan, the United States, Latin America, Saudi Arabia, and South Africa showed that 40 per cent of those who replied said they bought organic food, but of these, the majority (76 per cent) said they did so because they believed it was healthier and only 10 per cent mentioned any kind of disapproval of non-organic farming methods.[8] Animal health matters to people, even those who are indifferent to what happens to the animals themselves, because of what they see as its direct effects on their own lives. Their own interests are served by ensuring that any animal products they eat come from healthy animals.

Oddly enough, it is only quite recently that the links between animal health and human health have been systematically investigated and emphasized as a reason for improving animal welfare.

This is now a rapidly growing area of research.[9] There is increasing evidence that animals kept in conditions where their welfare is poor can have weakened immune systems and so be more likely to succumb to diseases.[10] Tom Humphrey has even coined the phrase 'happy chickens are safer chickens',[11] meaning that the best way to promote healthy foods to people is to promote good animal welfare. Health in animals is therefore not a luxury that we may or may not be able to afford, nor does its value have to be based on the gamble that animals may or may not be conscious. It is part and parcel of a multidisciplinary (what scientists say when they mean by 'holistic') effort at fighting human disease, along with microbiology, ecology, public health, and epidemiology. In the fight against human disease, the welfare of animals may turn out to be just as important, if not more so, than vaccinations, medication, and other more obvious treatments. It may even be that the reason such multidisciplinary studies are only just getting going is precisely because the case for good animal welfare has placed too much emphasis on the experiences of the animals themselves and not enough on the interactions between animal and human health. Animal welfare science, by seeing itself as making the case for animals based on its own definitions of 'good welfare', may have seriously missed a trick as far as actually implementing improvements in welfare is concerned.

Even where there is no direct evidence for effects on humans, the value of animal well-being to human well-being can be seen through other benefits such as the taste of food and whether people really want to eat it. The quality of meat, for example, is affected by how the animals have been kept and, in particular, how they have been transported. During the journey from farm to slaughter house, pigs lose weight, have elevated heart rates, and show a number of physiological changes such as dehydration and depleted levels of muscle glycogen. This results in dark, dry, tough meat that is less palatable.[12] People unconvinced by the idea of animal welfare for its own sake might well begin to take notice of it if it contributes to higher food quality.

This 'consciousness-free' view of animal welfare may be deeply upsetting and even offensive to those who believe that there really is something special about animals and that they are not just like plants or soils or rivers—just there for the exploitation of human beings. We depend on plants for our survival and so plant health is a desirable end for that reason. But animals are different. Animals are beings in their own right, not just our food or our playthings. Reducing them to the status of plants may seem to be a retrograde step in the history of animal welfare.

It is important to emphasize, however, that no such reduction is occurring. By linking animal health to human health and human well-being, animal welfare is given a powerful new set of arguments for why it is important. Nothing has been taken away. Those who believe that animals are conscious and should be treated ethically for that reason are still able to go on supporting animal welfare for that reason. It's just that some new recruits have joined the ranks. Those new recruits—the consciousness-free brigade—are bomb-proof in the face of attack by the killjoys, who try to argue that the evidence for animal consciousness is less impressive than had previously been claimed. They can therefore fend off attacks from sceptics that more traditional arguments may have difficulty with. They provide substantial reasons why animal welfare is important even to those who regard animals as ethically of no significance in their own right. What is even more important is that they might tip the balance for a large number of people who are uneasy about some of the things that are done to animals but are not quite uneasy enough to pay for things to be done differently. They might well be prepared to pay, however, if it also gave them safer, healthier food and also had environmental benefits.

Even people who are unconvinced that animals have conscious experiences exactly like ours may have some vague feeling that animals may feel 'something' without being able to say what. They feel more comfortable knowing that the animals have been treated well, without wanting to grant full human moral status. Roland Bonney sums up the attitude of many such people to the way farm animals are kept:[13]

> I have seen experienced commercial food company managers react violently to witnessing, at first hand, a 'conventional' intensive pig finishing unit—by denying that this is, in fact, how much of our pork is reared in the UK. It is interesting that very few people seem to want to go vegetarian. What they wish for is to be able to buy products from farming systems that do not make them feel ashamed.

Not feeling ashamed or, better, feeling good about the way animals are treated by society is an important ethical driver in its own right. 'A sense of well-being comes when people are able to consume food that is produced in a way that reflects their ethical aspirations,' Bonney writes,[14] and this applies particularly to food that comes from animals. It doesn't matter what the scientific evidence for consciousness is. People want to feel good about what happens to animals. If these vague feelings of wanting to feel good can be reinforced with what is good for their health and their diet, then the case for animal welfare will have a new set of voices in its support.

It is always easiest to persuade people to do something if they see it as being in their own self-interest.[15] Expecting people to act against the interests of their own children, for example, is extremely hard, even if it would greatly benefit other people's children. But show them something that benefits them, their children, *and* other people's children and the situation is transformed. So it is with benefiting animals. The 'and' is crucial. Reducing the risks of disease transmission and increasing the amount and healthiness of food are two utterly selfish human benefits that could provide the basis for good animal welfare.

There are, of course, many other benefits that humans derive from making sure animals are kept healthy besides just these two. The health of guide dogs is of direct importance for the mobility of their owners, and the health of draught animals is essential for the people who rely on their labour. The list goes on. You will be able to think of many other ways in which the health and well-being of animals directly or indirectly benefits human beings and could provide a reason why their health is important, even to people unconvinced by any arguments to do with animal sentience. You could believe that animals were food like plants and totally without feeling or sensation and yet the welfare of animals would still affect you. You cannot escape just by categorizing animals as without consciousness or by hiding behind the explanatory gap as a reason for saying that we cannot know for certain whether or not they are conscious. You could even assert that they almost certainly are not conscious because they do not have Higher Order

Thoughts (Chapter 5), but that would still leave you vulnerable to the next animal-based pandemic or food-borne *Campylobacter* infection.

If you are human, there is no escape from animal welfare. Almost all the goals that the human species is now setting itself—feeding a rising human population, reducing pollution and greenhouse gases, conserving habitats—depend either directly or indirectly on other species and their health. To believe that any of these goals could be achieved without taking into account the well-being of animals is to misunderstand our dependence on them. Their health affects our own health, our food, the medicines we have, and the ones we need, as well as making the earth a good place to live.[16] Not all animal welfare can be 'cashed' in terms of what it does to humans, but a lot of it can.

For far too long, animal welfare has been thought of as an isolated 'cause' with its own scientific meetings, its own campaign organizations, and its own journals. The right questions about its connections to human health, food production, and other issues have, until recently, simply not even been asked, either by those inside or by those outside the subject. But if, as seems likely, animal welfare turns out to be one of the best defences we have against diseases that affect humans, its whole status will have to change. Animal welfare will no longer be the poor relation, begging to be noticed. It will be there, centre stage, in discussions on the future of human food production, tipping the balance between healthy sustainable food production and disasters caused by a blinkered pursuit of efficiency at all costs.

But if animal welfare is to justify this new, high-profile role, it has to be solidly backed by scientific evidence and that means being clear about what 'animal welfare' actually is, as well as specifying exactly what would count as increasing it or decreasing it. So far in this chapter, animal welfare has been defined in the most basic possible terms—whether or not the animals are physically healthy. And 'healthy' has also been discussed in the most obvious and unequivocal of ways, such as whether the animals are infected with disease that they might pass on to humans.

But 'good welfare', to most people, means much more than just ensuring that animals are not dying of a disease or are so badly injured that their wounds are fatal. So what exactly is the 'more' that needs to be added to the definition of good animal welfare?

Basic physical health may not be everything we want to include but it is a good starting point. An animal that is injured or diseased is, like a plant that is injured or diseased, more likely to die than a healthy one. So for purely pragmatic reasons, humans have a vested interest in not letting them get to this point because healthy plants and animals are more likely to provide the benefits that humans need. This means that humans will benefit from being able to detect preclinical signs of ill health in their plants and animals and use these to take pre-emptive action. If they do something about the first signs of ill health or malfunctioning due to lack of water or nutrients, then the really damaging and possibly fatal later symptoms may never appear. The vital nutrient is supplied. The temperature is adjusted. Prevention is usually better and less expensive than cure. The skilled caretaker, whether of cows or tomatoes, will be able to spot the early warnings and step in before any damage is done. In this, he or she is similar to the skilled mechanic who detects an unevenness in the running of an engine and replaces a vital part before it breaks and damages the rest of the engine.

This very pragmatic view of animal welfare science sees many measures of 'welfare' simply as useful early warning signs that something is wrong with the animal's health. If nothing is done about them, the animal will sicken and die, so it is good practice not to keep animals in conditions where disease or injury may occur. This means that an animal doesn't have to be at death's door to have a potential health problem. It could be still quite 'healthy' but on the road to ill health.[17] Healthy animals are those whose bodies not only function well but are also without signs known to be associated with future malfunctioning. They also have signs that their bodies are functioning well, such as glossy fur or feathers, bright eyes, and plenty of energy.

By including absence of indicators of potential ill health as part of 'health', the definition of animal health has clearly expanded somewhat. The emphasis on whether an animal's body is functioning properly or is showing early signs of going wrong would imply that much of what we call animal welfare science is nothing more than preclinical veterinary medicine. Now, clearly, animal welfare science has much to offer vets in the early diagnosis of disease, although this, too, is an under-researched area where we need much more good evidence. But is that all there is to animal welfare? Do all signs of 'poor welfare' point inevitably to a decline into ill health and inevitable death? To understand why animal health is important, but not the be-all and end-all of animal welfare, we have to take into account a quite extraordinary feature of life on earth that enables it to survive in the face of the massive forces of destruction ranged against it. As this is a feature of plants as well as animals, we can still discuss animal welfare without necessarily invoking consciousness, although, as ever, consciousness refuses either to be ruled in or to be ruled out.

8

THE TWO PILLARS
OF ANIMAL WELFARE

The evolution of life can be seen as the evolution of the go-getters—that is, of organisms that found more and more complex ways of getting what they needed to stay alive. In the beginning were organisms that could only exist where conditions were right for them. But soon, the proactive ones were able to survive even if the conditions in which they found themselves were inadequate because they were able to go out and look for their own nutrients or light or whatever it was that was missing. Even bacteria can move up gradients to get what they need. Now, we humans take catering to our needs, present and future, to new heights. We plan ahead for what we might need next week, next year, and even for the rest of our lives. We take out insurance policies, plot, deceive, and form conspiracies. Many people save for years so that they have what they will need during a far distant retirement.

In between these two extremes of bacteria moving up a food gradient and the specifically human ways of planning for the future are countless other ways and devices used by other species to get what they need to stay alive and to reproduce—begging for food from a parent, stealing food from more successful competitors, spinning a web to trap food, or simply being so aggressive that all other organisms keep away from the food, the female, the nesting site, or whatever is needed for survival. Some of these ways are 'blind' instinct. Some are learnt. Some may be consciously planned. It is in understanding the mechanisms behind these different ways of getting what is needed for survival and reproduction that we see both the glory of life on earth and the tragedy of what happens when they go wrong.

The glory is the countless numbers of ways that organisms have found of meeting their needs and being able to stay alive—scavenging, eating each other, attracting mates, or trapping the sun's energy directly. The tragedy is the fact that most of them fail. Not only are most of the species that have ever existed on earth now extinct (over 99 per cent),[1] but most individuals of even successful species die before they have had a chance to reproduce. For every animal we see alive, there have been millions if not billions of rivals that didn't make it, whose bodies could not meet their own needs. Their mechanisms for getting energy to fuel their bodies failed because they could not, for some reason, find enough food. Or their mechanisms for keeping themselves at the right temperature failed because the environment forced them into a temperature zone that their heating or cooling mechanisms couldn't cope with. Or their needs came into conflict with the equally strong needs of another organism. Whether the lion seizes the throat of the zebra and pulls it, struggling, to the ground or the zebra swerves at the last moment and escapes—either way somebody's carefully evolved mechanisms for staying alive have let them down.

This pattern of the glory of life and the tragedy of its failure has been played out on earth for hundreds of millions of years, with no one around to see anything either 'good' or 'bad' about death, injury, disease, or reproductive failure. It was only when humans came along that any kind of ethical judgement on death or the processes of death was made. Most people do not pass judgement on the crocodile that kills its victim by drowning, or the long slow death that a cat might inflict on a mouse. But humans doing the same thing would be seen as cruel or as responsible for inflicting suffering. When hunting with dogs was outlawed in the UK, this was not a judgement on the ethical behaviour of dogs, but on the humans who took them hunting. With wild animals, death and suffering just happen. Where humans have some control or some influence, we start making ethical judgements about good and bad.

'Death', a veterinary surgeon once told me confidently, 'is not an ethical issue.' What she meant was that when an animal is dead it is no longer in pain; it is the suffering of living animals—what happens to them before they die—that matters. 'Putting it out of its misery' (i.e. killing it) is considered by many people to be the right and ethical thing to do to an animal (but, interestingly, not a human) because it puts an end to suffering. As far as animals are concerned, 'good welfare' is not generally taken to mean avoiding death, although some people disagree. It is about what happens to them before they die. It is the journey, not the destination, that counts. So, where along that journey from buoyant good health to close to death does good welfare become bad welfare? Bad health (injury and disease) are uncontroversial indicators that an animal is well along the road to death and destruction. Preclinical warning signs of impending death would also indicate that the journey has at least started. But what about the animal that looks healthy enough and exhibits no early warning signs of any known disease? Is its welfare necessarily good or might there be other reasons to believe that it might not be?

Living organisms—animals in particular and to a lesser extent plants—have two main mechanisms for overcoming the ever-present forces of destruction that constantly surround them. These are the ability to repair or restore themselves if they are damaged and—even more important than this—the ability to anticipate danger and so avoid being damaged in the first place.[2] These two mechanisms distinguish life from non-life. Non-living stones get worn down and damaged until eventually they are nothing but sand. A living plant or animal repairs the damage, heals the wound, fights off the disease. Of course the repair mechanisms do not always work and some wounds and some diseases are fatal. The fuel or the oxygen may run out. But the living (as opposed to the non-living) fights back. It tries. It does not go gently into any good death without putting up some sort of struggle. The struggle is apparent in almost every aspect of its structure, its physiology, and its behaviour. Its claws grapple with its attacker. Its immune system mounts an attack on the invading disease, blood clots its wounds, whilst behaviourally it remains hidden and immobile, helping the healing process to take place until it is 'whole' or 'healthy' again. Its skin regenerates, even the bones repair. To use Don Broom's words,[3] organisms have evolved a finely tuned ability to 'cope' with a wide range of threats and to restore themselves to healthy functioning when this is disturbed or damaged.

But the extraordinary thing about life is that it goes far beyond simply reacting to what has already happened to it. As life evolved, it rose above merely coping or repairing itself when damage has been done and we now see a whole raft of mechanisms that enable living organisms to be proactive and to take steps to anticipate and avoid threats to survival long before they have any effect at all. For example, instead of waiting until a predator has actually injured them and relying on their healing mechanisms to get them through the crisis, animals take precautionary measures long before the predator has even appeared over the horizon. They only feed at certain times of day when the predators are less likely to be about. They have acute senses of smell and hearing to warn them of predators approaching. They group together in flocks or herds, building social bonds for days, weeks, or even years before the safety of the group saves an individual from death by warning it of an approaching predator. It is not only eyes, ears, and noses that are ways of anticipating and therefore of avoiding injury altogether. An attraction to other animals of the same species and a tendency to keep close to them are as much part of not dying from a predator's bite as are tough skins, clotting blood, or a sharp turn of speed. It is just that they operate further away in time from the point of death.

This ability of animals to anticipate danger before their survival is actually at risk is possible for exactly the same reason that scientists are able to make sense of the world. This is that the world is at least a partially predictable place. Night follows day. The seasons progress in an orderly fashion. The colour of a fruit predicts its flavour and its food value. Animals, like scientists, use these predictable sequences to anticipate what might happen next, what danger might be nearby, what might be worth investigating. Predictable patterns in nature are a wonderful thing. Use them and you can be ahead of the game. Forecast and you can be much better prepared. Anticipate and you might save your life.

And that is what both animals and plants have done. They have cheated death by anticipating the guise in which it is most likely to occur and then getting out of the way before it actually strikes. They start eating before their food reserves fall dangerously low and long before they are in any real danger of dying of starvation. They hoard food in times of plenty to anticipate the hard times. Even the seeds of plants 'predict' the best time of year to grow, many of them requiring a period of being cold and even frozen before they will germinate. They make use of the fact that winter predictably comes before spring to avoid germinating at the wrong time of year and being destroyed by frost.

Everywhere we look, we see anticipation as a powerful way of avoiding death and destruction. The fact that seeds do it shows that this does not have to be done by a complex brain working out the best days of the year to do something. 'Anticipation', in the sense of using one event as a predictor of another, can be done automatically and quite unconsciously, and need be no more mysterious than your car flashing a light to warn you that you will soon run out of fuel. The car uses a low level of fuel to predict that, unless you fill up, the car will stop because it needs fuel. (Incidentally, even cars 'anticipate'. They don't wait until you have actually run out of fuel. They are fitted with fuel gauges and inform you well ahead of time that, unless you fill up, you will stop. As a result of this built-in anticipatory device, you will have a much more efficient journey than if you were constantly breaking down and having to walk to the nearest fuel station.)

Sometimes the anticipation mechanisms that animals use are as directly related to a future event as a low level of fuel in a fuel tank is to actually stopping altogether. For example, when we become thirsty, our body's need for water is signalled by special cells in the brain that have themselves become dehydrated.[4] These dehydrated cells are part of a mechanism that leads us to look for water and drink. The cells provide a direct connection between the need for water and the drinking behaviour that can fulfil that need. The body is effectively able to predict how likely death by dehydration is to occur from the remaining water reserves and then to make us 'thirsty' in proportion.

But sometimes the prediction of a future event rests on a much more complicated set of events, much more distantly connected in time. For example, many species of birds, such as swallows and martins, breed in the northern hemisphere but fly south in the winter to avoid the low temperatures and food shortages of a northern winter. But they don't wait until they experience a lack of food or cold temperatures before they migrate. They leave their summer quarters long before things get tough and long before they start to feel cold or find it difficult to find food.[5] They make use of the fact that towards the end of every year, the days start to get shorter. Shorter days and longer nights are good predictors that winter is coming, even when the balmy days of early autumn are still warm and buzzing with insect food. The birds do not have to realize in any conscious way that shorter days mean that winter is coming. The shorter day lengths and their own internal body rhythms simply drive them to change their behaviour. They no longer build nests or rear young as they have done all summer. They eat to put on weight[6] so as to have enough fuel for the long journey south, again without any need to 'realize' what they are doing. And then they are gone. A few weeks later, the cold sets in but they never feel it. They are (if they have survived the journey) far away where it is not cold and where there is plenty of food.

Now this separation in time between the hazard the migratory birds are anticipating (the cold winter that could kill them) and the cue they actually respond to (the shortening day length in the autumn) has some interesting consequences for animal welfare and shows why just being concerned with an animal's health can never tell us all we want to know about its welfare. Imagine a species of bird that normally migrates, such as a warbler. Now imagine a kindly human being trapping some warblers in the late summer and putting them in an aviary. The aviary is large, provides protection from predators, shelter, and plenty of food. Everything is done to ensure the health of the warblers, but there is one problem. The warblers can fly within the aviary but they cannot fly free and they cannot therefore migrate. The kindly human insists that the health of the birds will be outstanding all during the winter—much better than that of wild birds because they will be protected from all the hazards that they would normally have to fly south to avoid. In fact, their chances of survival to the following spring are a great deal better than those of wild warblers facing a difficult journey of many miles, predators, humans with guns, and unpredictable weather. They have no 'need' to migrate in the sense that they will not die if they don't migrate, so migration is not like a 'need' for food where animals do die if they don't get it. But the behaviour of the birds suggests that, despite having everything they need for survival, their welfare is not good. Take a look at what the birds do.

The birds show what Germans call *Zugunruhe* or 'migratory restlessness'. They attempt to fly but are prevented by the wire mesh of the aviary. They incessantly hop from perch to perch and then flutter against the wire, looking for a way out. They appear to be agitated and flighty. The shortening days have stimulated them to migrate, even though the conditions in the cage are good and there is plenty of food. It is impossible to explain to them that they have no need to migrate because there will be food and warmth throughout the winter and they are much safer where they are than out there in a world of hawks and blizzards and people with guns. They may have no need to migrate but they want to migrate—desperately. They are highly motivated to fly because, in nature, flying south for the winter, however hazardous the journey, is the best way to avoid the even greater hazards of staying put and dying of cold or hunger. But in the unusual conditions of an aviary, where humans have interfered, the 'need' and the 'want' have become separated.

By breaking the natural connection between shortening day length and worsening living conditions, we have also broken the connection between wanting and needing. The birds now want something they do not need for their survival. However much we cater for their physical needs—food, shelter, water—so that their physical health is ensured, they still want to migrate. Their ancestors—the ones that successfully survived and reproduced in the past—did so because they evolved the motivation to migrate when days became shorter and nights started to draw in. Their descendants—the ones we put in a brand new environment—are confronted with a world where none of the previous contingencies hold and are in many cases reversed.

Plenty of food will be available all winter. Not flying south could be better for health and survival than migrating. It is a world where the old certainties are gone and the connections between events that have been true for hundreds or millions of years no longer hold. Natural selection did not prepare animals for this brave new world in which human beings, not nature, decide what consequences follow from what events.

The problems with animal welfare arise because animals cannot always adapt to these new eventualities, particularly the ones that humans have devised for them in zoos, farms, or laboratories. They are left with a legacy of how to live in an old world, the world of their ancestors, with rules of how to behave hard-wired into them, based on the old ways and the old dispensations. Sometimes they do adapt, of course. Even some wild-caught animals can be tamed. Rats, mice, sparrows, and numerous other animals have adapted to the presence of humans, seeing us as a bonanza of food and an incomparable form of shelter. Domestication[7] has also changed many species so that they have genetically adapted to life with us in a way that parallels the ways their distant ancestors genetically adapted to the wild through natural selection. But even for the animals that we have selected and moulded to our own use, even those that depend on us for their food and shelter, even they often have a hard core of 'wants' that is their legacy from the past and may not be satisfied in zoos, farms, laboratories, or even the homes in which people keep them as pets. They may 'want' to perch, or migrate, or dig, or search for their own food even though they have no need (for survival) to do these behaviours because we humans have taken care of them and made sure they have everything they need to keep them healthy.

So we can now see that 'good welfare' for animals is more than simply making sure that their physical needs are taken care of and that they are in good health. They 'need' food, water, and shelter, certainly, or their health will deteriorate and they will die. They are also brought closer to death by being infected with a disease or injured so if, for purely pragmatic reasons, we want to keep them alive, we attend first to their physical health, just as we do with plants. But we can also see that if all we did was to attend to their physical health, we would be overlooking some of the most important things about them, the very things that make them into the species they are. We would leave the warbler fluttering against the wire walls of the aviary and the polar bear pacing endlessly round its concrete enclosure. Animals carry with them a legacy from their ancestral past. If we put them in environments that are very different from the ones in which their ancestors evolved, that legacy will make them want to do things they no longer have any need to do.

In fact, animals can 'want' to do the things they no longer need to do just as strongly, if not more so, than the things they really do need to do. For example, starlings will choose to forage for food using the natural behaviour of pecking in grass in preference to taking all their food ready prepared from a dish, even though the natural foraging takes time and effort and actually gives them food at a slower rate than feeding directly from the dish.[8] If the need for food were all that mattered to them, we would expect them to eat exclusively from the dish, where they can get food 'for free'. But they don't. They 'want' to go through all the time-wasting, troublesome business of obtaining exactly the same food by natural means. Wanting to do the behaviour normally associated with getting food is an even stronger instinct than their need to get food as quickly as possible.

We see the same phenomenon in dogs endlessly chasing balls. Chasing small moving objects is clearly part of the natural hunting behaviour of a carnivore such as a dog, but even a well-fed dog that doesn't 'need' to hunt to stay alive still 'wants' to do the chasing part of the hunting sequence that, in its wild ancestors, would have been a necessary precursor to fulfilling the need for food. The 'need' (for food) and the 'want' (to chase) have, in domestic dogs, become so separated that dogs will 'hunt' balls for their own sake, with no food reward in sight. Trainers of sniffer dogs often use 'playing with a ball', not food, as the dogs' reward for finding particular substances. The dog 'wants' to play and chase so much that this has itself become a major motivator of dog behaviour, even though there is no need to do so to obtain food.[9] Good welfare therefore has to cater not only for what animals need to keep them alive but also for what they want.

To find the best way of doing this, we need to go back to the opening theme of this chapter—the complexity of ways in which animals fulfil their basic needs in nature and the fact that many of these ways involve responding to events that may be only indirectly related to their health and may be quite widely separated from it in time. We need a word that takes into account the proactive, predictive ways in which animals have evolved in order to get what they need or to get away from danger. The word must be able to encompass both the desperate fluttering of the caged migratory bird and the obsessional ball chasing of pet dogs. It has to be a word that takes into account what ultimately happens to a wild animal in its natural habitat, but it must primarily focus our attention on what it is, proximately, in the here and now, an individual animal is actually responding to. It has to be a word that allows us to make the distinction between, say, needing food and finding the taste of it so pleasant that we can't help eating it even if we don't 'need' it. Biologists have a useful but somewhat pedantic way of expressing this distinction between the ultimate, evolutionary reason why something evolved and the proximate, immediate mechanism that triggers it. But rather than talking about ultimate needs and proximate needs, it seems much more straightforward to refer simply to 'needs' and 'wants'.

Needs (primary needs) can then be related directly and easily to health, life, and death, in both present-day animals and their wild ancestors. Wild animals that did not meet these needs died or failed to reproduce and were eliminated by natural selection. Captive or domesticated animals whose needs are not met by humans will also die. 'Wants', on the other hand, are part of the proximate mechanisms that animals have evolved for fulfilling those needs in practice. Some 'wants' can obviously and easily be related to needs. We both need food and 'want' it. The feelings of hunger and the pleasantness of the taste of certain foods are part of our mechanism for refuelling our bodies. But some 'wants' have become divorced from their original needs so that the well-fed dog (or possibly even the well-fed human) still 'wants' to perform at least some of the hunting movements that, in the wild, helped its ancestors to fulfil their need for food.

It is very important to realize that this divorcing of wants and needs is not just a function of domestication or the artificial environments in which we humans put animals (and indeed ourselves). Although artificial environments, or any environment that is different from that in which the ancestors evolved, may show up the distinction in particularly sharp relief, the divorce between the two is there in wild animals too. It is part and parcel of the diversity of ways that animals have found of getting their needs met and maintaining good health. It is part of the glory of the diversity of ways in which different species make their livings. Wants are one of the most dramatic ways in which all animals, particularly wild ones, get their needs met and the way they avoid injury. They are the servants of needs, part of the many mechanisms by which bodies keep themselves alive. Wants and needs are not the same. It's just that in wild animals, we may not realize that the distinction is there.

The Dutch ethologist Niko Tinbergen did more than almost anyone to show how the immediate stimulus that triggers behaviour can be separated from its evolutionary origins—or, in other words, how the proximate mechanism of what the animal wants to do can be separated from the ultimate evolutionary function of what it needs to do to survive and reproduce. Tinbergen famously observed that some male sticklebacks he had in a tank on the windowsill of his laboratory suddenly became extremely excited by the arrival of a red mail van outside.[10] Now, male sticklebacks become bright red underneath in the breeding season and vigorously defend their nests against other males, which are also bright red. The sticklebacks had evidently mistaken the red mail van for a rival male stickleback. A very elementary mistake, one would have thought. However, the 'mistake' that showed up so dramatically in the laboratory would never have happened in the wild. In the streams where sticklebacks live, a bright red object is, almost certainly, another male stickleback because there just aren't any other red objects around. So other male sticklebacks can be recognized with the simplest of cues—redness—without any need to check out the number of fins, or the details of its shape. The rule of thumb 'if it's red and roughly the right size (which even a mail van would be at a distance) then attack it' is normally a perfectly good, if crude, mechanism for putting the adaptive advantage of chasing away rivals into practice. There is, normally, no need to have any more complex mechanism than this. It is only in the highly unusual environment where a mail van comes into view that the simple rule 'fails'.

Just in case you are beginning to feel superior to animals that do such 'silly' things as try to attack mail vans or try to sit on ostrich eggs (another trick that Tinbergen played on unsuspecting gulls and oystercatchers by putting these in their nests), just remember that we humans are not beyond doing something very similar. Melissa Bateson[11] put a photograph of a pair of eyes beside an honesty box, where people were supposed to pay if they made themselves a cup of tea or coffee. She found that people were far more likely to pay up when the eyes were there than if there were a photograph of something else. No face, just a photo of a pair of eyes was enough to change people's behaviour and make them feel they were being watched. A very simple stimulus, with a big effect. It normally works perfectly well because we evolved in an environment where eyes meant someone was indeed watching you. In this case, there is no need to respond to a photograph of eyes because in fact there is no one there. But people still want to behave differently in the presence of the photograph because responding to eye-like stimuli is part of the mechanism by which people interact socially.

Tinbergen's distinction between the adaptive reason as to why something evolves and the actual mechanism that guides the behaviour of the animal (or person) on a moment-to-moment basis is of crucial importance to animal welfare and to formulating both what we mean by it and how we might measure it. In particular, it shows us that what animals want may not always be what they actually need to stay alive because the evolutionary advantage of doing something (obtaining more food, avoiding predation) may be connected in a very complex and indirect way to the mechanisms by which they have evolved to obtain it. As the go-getters evolved, natural selection favoured those that could anticipate danger or their own future needs by wanting things before they needed them and responding to clues as to what the future might bring. Any formulation of what is meant by 'animal welfare' thus has to take into account both the long-term needs and the short-term wants that have evolved in wild animals and are still the legacy of captive ones.

Here, then, is a simple working definition of what we might mean by 'good welfare'. Good welfare is the state of an animal that is both healthy and has what it wants.[12] In other words, both its needs and its wants are met. This does not mean that all its individual needs and wants are perfectly met all the time. As Robert Hinde pointed out many years ago,[13] most wild animals are in a permanent state of conflict because satisfying the need for water by going to a waterhole will put them in danger of predators. Running from the predator will mean an unsatisfied need for water or food. Most animals are a bit hungry, a bit thirsty, a bit fearful, a bit too hot, or a bit too cold for most of the time. What natural selection has done is not to produce animals that live permanently in a state where their needs and wants are perfectly met, but animals that achieve the optimal compromise between satisfying their different needs and wants in that environment.[14] Indeed, the very fact that wild animals do live in a state of conflict should make us extremely wary of going too far in satisfying animals' individual needs and wants. If we give an animal all the food it wants all the time so that it is never hungry, it may well become obese and unhealthy because in nature, natural scarcity of food would mean that it would sometimes be hungry and would therefore not over-eat. There is an optimal level of meeting both wants and needs.

This two-pillared working definition of good welfare has the advantage that it can be readily understood by everyone—biologists, non-biologists, people who know a lot about animals, people who know very little, people who are convinced animals are conscious, people who are sure they are not, politicians, pet-owners, and that ubiquitous category 'consumers'. Everyone. It has the even greater advantage that although everyone can understand it, it also points directly to what any scientific approach to animal welfare has to do. It says that we have to find out what makes animals healthy and we have to find out what the animals themselves want. Anyone claiming that animal welfare can be improved by, say, giving animals more space or providing an 'enrichment', now has to demonstrate either that the claimed improvement actually improves the animal's health in some way or that it provides the animal with what it demonstrably wants, or both. An enrichment that had no effect on an animal's health and was ignored by the animal itself could not be claimed to improve welfare, however much better it made well-meaning human onlookers feel. The two questions together provide the basis for evidence-based animal welfare.

But although the two-question approach to animal welfare is simple, it is far from simplistic. It handles, as a delicate balancing act between the two pillars, the separation between what the animal needs for good health and what it wants. It leaves open the possibility that animals (and people) do not necessarily always choose what is best for their health in the long run. It says explicitly that asking one or other question on its own could be extremely misleading. We have to understand both health needs and wants, not one or the other. That is why there are two pillars to animal welfare, and why two questions, not just one, have to be asked. Certainly the two questions may give contradictory answers. Your dog may want more food but giving it to him may make him obese and unhealthy. His health needs and his wants conflict and you will need to make an evaluation of what to do. But resolving such conflicts is no more problematic than the fact that your child doesn't want to go to the doctor but needs to for health reasons. Or the fact that you don't want to have an operation but need surgery to have any chance of survival. As we have seen all along, wants and needs sometimes collide and sorting them out may be a difficult matter in any circumstances. At least the two-question definition shows you clearly what your choices are. The dilemma is starkly set out in front of you rather than being lost in a cloud of excuses that 'it's all very complex'. It's actually rather simple.

Superficially, defining animal welfare in terms of just two simple questions appears to fly in the face of much of what has been written and said about animal welfare over the last thirty years. It has become almost mandatory to say that good animal welfare is multifactorial and that we must take into account as many different measures as possible in order to arrive at a true picture of welfare. The European Union Welfare Quality report starts with the words 'Animal welfare is multifactorial' and continues,[15] 'It is now widely accepted that animal welfare is very complex …' It then proceeds to recommend taking thirty to fifty measurements of welfare for each species of farm animals. Almost everyone who has written about animal welfare says the same thing. Animal welfare is complex, not simple; and it needs many different measures.[16] The prevailing view is that if we take lots and lots of measures then somehow, out of a long list of different measures, a consensus will emerge (it is usually not clear how). New 'measures' of welfare, such as body temperature, immune depression, fractals, sleep, play, are constantly being added to the list.[17]

So reducing the number of questions we need to ask to establish good or bad welfare down to just two would therefore seem to be bucking the trend and taking us in a completely different direction from where the rest of animal welfare science is going. In fact, the opposite is true. The many different measures that people take on 'welfare' can all be seen as addressing exactly these two questions. For example, measurements of longevity[18] and lameness[19] address the question of how healthy the animal is directly, while measurements of immune function,[20] abnormal hormone levels,[21] and disturbed behaviour[22] address it indirectly through providing early warnings of health problems to come. Choice tests and measures of motivation are direct ways of finding out what animals want, as we will see in more detail in Chapter 9, while measurements of frustration, deprivation, and boredom are all indications that an animal does not have what it wants.[23] Measures of 'positive emotions' can be seen as the opposite—measures of what happens when animals do have what they want.[24] The two questions thus unify what superficially appear to be very different approaches to measuring welfare and make it clear that most of them are addressing the same two issues.

We can see how this works in practice by applying the two-question approach to one controversial approach about what constitutes good welfare—namely, the widespread belief that welfare can be assessed by how closely the behaviour of captive animals corresponds to the 'natural behaviour' of wild animals. Natural behaviour, according to this view, is a sign of good welfare.[25] By definition, natural is good. The assumption that the more natural the behaviour of animals the better their welfare is behind many enrichment programmes in zoos, where enrichments are often specifically designed to increase the opportunities animals have for behaving naturally. The Five Freedoms (a widely used list of what good welfare is for farm animals),[26] which include freedom from hunger, thirst, and discomfort, also state categorically that animals should have the freedom to 'perform most normal patterns of behaviour'. But how valid is this assumption? Should 'natural' be added to the requirements for good welfare?

By asking the two questions, we can actually see what would provide a good answer. First, the health question. We can compare animals with and without the opportunity to perform natural behaviour, perhaps by providing some of them with various sorts of enrichment. Which is healthier (as judged by longevity, disease levels, and other veterinary measures of health)? Then, do animals want to behave naturally? Do captive animals want to do all the things that their wild counterparts do, or do they find plentiful food without having to hunt for it far more preferable? The connection between 'natural' and 'good' welfare becomes something that has to be established with facts by looking at the animals themselves, not just by making romantic assumptions about what life in the wild might be like. The results could turn out to favour either natural behaviour or not and there might be some surprises. Being chased by a predator is a 'natural' part of the lives of many wild animals, as are periods of food shortage. Would having the natural behaviour of being chased by a predator (which would anyway be illegal in many countries) improve the health of the animals? And is it something they would want? Many wild animals, from fish to antelope, do approach their predators and 'inspect' them at close quarters, often putting themselves at risk,[27] so this is not quite the outrageous suggestion it might seem. Periods of food deprivation—mimicking lean periods when wild animals do not have what they want—might actually improve health. The point is that the answer will depend on specific kinds of evidence, not a priori assumptions about which is better. It may turn out that what is 'natural' really does improve welfare according to the answers to the two questions. But the improvement in welfare is not just because the behaviour is 'natural' and natural is not good by definition, but because naturalness has been demonstrably linked to real evidence in the form of the two pillars of animal welfare.

Many leading writers in the field of animal welfare have definitions of welfare that are compatible with the key importance of the two pillars, although they might not use quite these words. John Webster, for example, talks about the welfare of an animal as being determined by 'its capacity to avoid suffering and sustain fitness'.[28] And David Fraser lists the three key components as basic health, affective state, and the ability to lead the kind of life for which they are adapted.[29] All the 'two-question' approach does is to simplify, and put into a slightly catchier form, what most people have been saying for a long time. The point is, this is not something radically new or imposed from outside animal welfare science, but a kind of distillation of what many people inside animal welfare science have been saying for a long time.

Bringing together what scientists mean by 'good welfare' with what non-scientists mean and can easily understand is an important part of building a consensus view on what constitutes good animal welfare, which, in turn, will help to ensure that animal welfare is taken more seriously in future. The continuing arguments about what is meant by good welfare and the complexity of some of the proposed measures give the impression of a lack of rigour in the evidence as well as making it difficult to provide practical solutions. Unrigorous, impractical proposals for animal welfare can all too easily be dismissed by those who don't think animals are all that important anyway. To find a definition of good welfare that everyone, whatever their views about animals, can buy into shows in a practical way what needs to be done to improve animal welfare.

The emphasis on animal health as one pillar of animal welfare makes it easy for animal welfare to be related directly to human health and well-being. Good animal welfare gets support from being part of the agenda for benefiting humans. At the same time, stressing the second pillar by also giving prominence to the question 'does the animal have what it wants?' draws in those who feel there is more to good welfare than just physical health, but without having to answer the hard problem first. The two-pillar approach shows how studies of animal welfare can be successfully carried out without any reference to the 'hard problem' whatsoever. You can choose to believe or choose not to believe that the animals in question have conscious experiences. If you do believe it, you have an added incentive for thinking that their welfare matters. But even if you choose not to believe it, you still have plenty of reasons to give animal welfare high priority—reasons for populating the world with healthy animals that have well-functioning bodies and strong immune systems instead of populating it with sick ones. The two questions point us firmly in the direction of the (relatively) 'easy' problems we still have to answer—about animal health and animal wants—while not letting us get distracted by trying to answer the really 'hard' ones about animal consciousness. They give us animal welfare, but without a whiff of consciousness in sight.

Or do they? Asking about what makes animals healthy is a practical down to earth sort of question with no overtones of consciousness. But asking about what animals want? Doesn't this bring us back to within striking distance of the questions about animal consciousness? Doesn't letting 'wanting' in through the door as the other half of animal welfare bring with it all the same problems and objections as attempting to study consciousness brings? Are we unable to escape it after all? Can there be a scientific study of what animals want?

9

WHAT ANIMALS WANT

Even with other humans, words are not top of our list for finding out what they want or how much they want it. We say things like 'He puts his money where his mouth is' or 'They voted with their feet' or even—as a complete gift to anyone trying to find out what animals want—'Actions speak louder than words'. We have ways of pouring scorn on words as unreliable carriers of true intention or real motivation. 'Mouthing', we say, or, as a final insult, 'lip service'. Words are cheap. Anyone can say things. It's what they do that really impresses us.

What people do can be anything from millions of people all turning on their televisions to watch a particular programme (showing how many people want to watch it) to paying a lot of extra money for convenience food (showing that they don't want the bother of cooking for themselves). We know that people want to watch football or baseball or the Olympics by the crowds that turn up. We know that they want to get a ticket for Wimbledon or a sight of the Royal Wedding by the fact that they are prepared to sleep outside on the pavement the night before. We are particularly impressed by people who are prepared to resign their jobs on a matter of principle or lay down their lives for what they believe in. Even if we don't speak the same language, we can judge what such people want and the strength of what they want by their behaviour.

The same idea of using actions rather than words can be adapted very easily to animals and to understanding what they want. The ecologist David Lack long ago identified what he called the 'psychological' factors that affect where different bird species are to be found.[1] He explained the reason why some species are found in woodland and others in open fields or on water by the fact that birds fly widely over different areas and then decide to settle down or move on depending whether a place has what they want. Particular sorts of trees, nesting holes, or open water might be some of the things birds would look for. Gordon Orians similarly talked about birds being 'turned on' by environments in which their chances of surviving and breeding successfully would be greatest.[2]

Modern technology has now given us the ability to study these natural manifestations of what animals want in much more detail. By fitting wild elephants with radio collars, for instance, it is possible to track their movements continuously as they move around their natural habitat. What they want is revealed not only in gross outline (the general area they want to be in) but in the minutest detail of where they want to be on a moment to moment basis.

Elephants, even bull elephants that were thought to be largely solitary for much of their lives, like to keep within rumbling distance of each other, keeping in contact by deep sounds that travel over a considerable distance. But, in addition, it has now been discovered that elephants have some quite specific and unexpected dislikes. Elephants do not want to walk up hills if they can possibly help it.[3] This can be shown by plotting a map that shows both where there is good elephant food in the form of vegetation and also the contours of the land revealing the hills and valleys. If the tracks of where the elephants choose to walk, as gathered from the radio-tracking data, are superimposed on this map it becomes clear that the elephants are keeping to the flat parts of the terrain. They avoid even small hills, despite having to walk further along the flat to get to the next lot of food.

The reason that elephants don't like hills is that, with their huge bulk and vegetarian diet, they have to eat enormous quantities of food each day just to keep going. Driving that huge bulk up a hill would significantly increase the amount of food an elephant would have to eat. For every 100 metres climbed, it would appear that an elephant would have to eat for another half an hour. No wonder they do not want to go up hills.

What animals want—as measured by where they go, what they avoid, and what resources they utilize in the wild—has now become an important tool for both conservation and for zoos. If a conservation area is being set up, it is important to know such things as how much space the animals want, what sort of nesting areas they choose, and what sort of distances they want to keep from other members of their species. The methods of establishing these are getting better and cheaper all the time. Bulky radio collars are increasingly being replaced by tiny devices that measure small body movements, heart rate, and body temperature 'on board', and are now so small that they can be fitted on small mammals, fish, and birds. Cameras can be fitted to birds to show where they fly and where they settle.

Good old-fashioned observation still has its place, too. With whale-watching becoming increasingly popular, many people have become worried about the effect that tourists are having on whales and dolphins themselves. Do they want tourists around? Some dolphins off the coast of New Zealand appeared not to. Tourist boats frequently visited the areas where the dolphins were feeding, to maximize the chances of their passengers being able to catch sight of them. However, the dolphins were often observed to stop feeding and leave the area when the boats arrived.[4] By observing the dolphins and timing the arrival of the various tourist boats, it was possible to show that the dolphins would tolerate the occasional tourist boat, but if more than one boat an hour came along, they would leave the area: a very clear indication that they did not want too many tourists in their feeding area.

Finding out what animals want by seeing where they go naturally and what they choose to avoid is, of course, no different in principle from finding out what plants want by looking at where they are found growing. Plant ecologists use just the same methods as bird ecologists and elephant ecologists in plotting maps of where plants are to be found in relation to such things as the acidity of the soil, moisture, or shade. Both plants and animals have habitat preferences more or less linked to what is good for their survival.[5] Plants 'choose' by growing well or badly, or by spreading their seeds that either germinate or not, whereas animals have, in addition, the dynamic short-term ability to choose something different every hour, every minute, or even every second. We call that behaviour. Animals are like plants that have the ability to keep changing their minds. But even here, with 'choice' and 'wanting' and 'liking', and now even 'minds', we don't have to invoke consciousness at all. The fact that it is possible to use essentially the same methods to study what plants want and what animals want shows that there is no need to do so. A good gardener will give his or her plants what they want, and will use information about where they choose to grow in the wild as the best source of information about what they might want. But the proof of whether the plant has what it wants is essentially whether it has what it needs to grow well. Animals have wants as well as needs as we saw in Chapter 8; what animals choose may not always be best for their health and well-being in the long run, particularly when they are in the odd and often unnatural environments we often keep them in.

People have therefore turned to other ways of understanding what animals want, ways in which they have rather more control than letting animals go about their natural business and seeing what they do. Sometimes the simplest of methods can give the most dramatic results. When chickens or female junglefowl incubate their eggs, they undertake an entirely solitary twenty-one-day vigil, as the males take no part whatsoever in this process. Once they start sitting on eggs ('going broody' as we call it), they remain stubbornly on the nest until the chicks hatch. While they are broody, the hens occasionally take brief breaks for food and water but these few minutes off the nest do not allow them to eat or drink nearly as much as they would normally do. Not surprisingly, they lose a great deal of weight while they are incubating their eggs.

They can lose as much as 17–20 per cent of their total body weight.[6] This is a massive weight loss. If a human being starved an animal to this extent, they would probably be accused of great cruelty. It would seem as though the hen must be extremely hungry. She must want food very much.

However, this is not what she wants at all. If an incubating hen is offered food and drink at the nest, so that she does not have to leave her eggs, she refuses them. She appears not to be hungry despite her massive weight loss that, in other circumstances, would point to extreme hunger. As a result of going broody, her whole body's metabolism changes so that the normal hunger mechanisms do not kick in even when her weight drops to what would, under normal circumstances, be regarded as a serious threat to survival. Here is a dramatic separation between what the hen needs and what she wants, a confirmation that animal welfare has two components, not one. She 'needs' food in that her body is undergoing a weight loss that normally triggers feeding. Indeed, this need for food is shown subsequently after the chicks hatch and she compensates by eating an extra amount so that she quickly returns to her normal body weight. But she temporarily doesn't 'want' food. Her refusal to eat even when food is offered is as eloquent a statement of what she wants and does not want as any words could be.

What animals want can also be discovered in a variety of different, often just as dramatic, ways. One obvious approach is simply to offer animals a straight choice between two objects, two foods, or even two environments and see which one they go for, as an indication of which one they most want. We can ask cows whether they want to graze out in pastures, for example, or whether they are content to be given concentrated nutritious food indoors. This is an important welfare question because cattle are grazing animals but are often housed indoors. Some people would argue that as long as they are given proper food, they have everything they need. Others would argue that being outside and eating grass is so much more natural that cows want to do this anyway, however nutritious their diets. Which is correct? In an attempt to find out what the animals themselves wanted, some high yielding dairy cows were offered a choice.[7] Twice a day, after milking, they were taken to a choice point from where they could see two possibilities. About 50 metres away in one direction was a pasture full of grass. The same distance away in the other direction was a house where a standard mixed ration was freely available. Having made one choice, they were then free to move to the other if they wanted to. From the choice point, the cows chose to go inside almost twice as often as they chose the pasture and they also spent more total time indoors. They also, not surprisingly, chose to spend more time indoors when it was raining. This study demonstrates both the potential of, and some of the problems with, very simple choice tests. On the one hand, the cows were, quite literally, voting with their feet. They were expressing their preferences both immediately (first choice) and over the longer term (total time spent in one or the other environment). But of course, it could be argued that the choice was not quite 'fair'. There was a concentrated food ration inside and, as high-yielding dairy cows, they may have needed this to meet their nutritional demands. Grass outside, however attractive, may not have given them enough for their nutritional needs so they may have been choosing to go indoors because of the food rather than because they 'liked' the environment. Certainly this conclusion is made more likely by the fact that the highest-yielding cows were also the ones that spent the most time indoors. So perhaps a fairer choice that didn't involve having to choose between food and being outside quite so starkly would have given a different result and shown that the cows did want to be outside. It's a question of designing the right experiment to give the answers that are most useful. The point is that while we may not yet know everything there is to know about what animals want, we have the means to find out. It is not a closed book.

On the other hand, choices in nature aren't 'fair' either. The only source of drinking water may be a river where crocodiles are likely to hang out, and the best nest sites may already be occupied by an aggressive competitor. Elephants have to choose between food and walking up hills. So in setting up 'choice' experiments for animals, even where the choices might be described as 'unfair', all we are doing is asking animals to do something that they have been evolved to do in the wild. We are simply putting a finger on the pulse of decision-making mechanisms that wild animals have evolved over millions of years. By studying their decisions or choices, we can find out where they are on the spectrum between being satisfied and unsatisfied, between having what they need to 'cope' and having a whole raft of unmet wants. Sometimes, as in the case of the little boy who was asked (by some unkind adult) whether he would rather have an ice-cream or his mother, such choices may sometimes be artificial and 'unfair', but we are increasingly seeing the use of natural choices and natural happenings to study what animals want. This gives a real feel not just for what animals want but also for what they want most—and least—of all.

For example, laboratory mice prefer a solid floor to a wire grid floor if given a straight choice and, not surprisingly, use a nest box if one is provided. However, if nesting material in the form of paper towel or tissues is provided only on the grid floor away from a nest box, the preferences switch and they spend the majority of their time with the bedding material despite it being on a wire floor.[8] The mice clearly want nesting material even more than they want a solid floor. Broiler chickens will jump across a high barrier to get away from a crowded area to one where there are fewer chickens,[9] indicating that they want to be where it is less crowded. This is made more impressive by the fact that they are normally reluctant to jump a high barrier to get at food unless they have been food-deprived for a considerable time, such as six hours. Ducks will similarly jump high barriers to get access to bathing water[10] and demonstrably want open water they can splash in more than they want other forms of water;[11] mink will push open doors with heavy weights attached in order to be able to swim in a bath of water.[12] All these examples suggest that animals want certain things badly enough to be able to go to some lengths and expend a great deal of effort in order to get what they want.

The idea that finding out what animals themselves want and then using this information to design better conditions for them is one that is increasingly being implemented in farms, zoos, and laboratory environments.[13] Of course, animals can't necessarily have everything they want in practice, any more than the rest of us can. It may be too expensive or it may conflict with other wants. And there will be times when what they want is not good for their health in the long run, just as with ourselves. But as the second (and, to stress this point yet again, not the only) pillar of animal welfare, what the animal wants is increasingly recognized as an important area of research in its own right.

An important step forward in this field has been the development of easy-to-use methods of getting animals to tell us what they want when we can be certain that the animals are already experienced in what they are choosing. It is one thing for an animal to be attracted to something (or to be repelled by it) the first time it encounters it. An unfamiliar food or an unfamiliar environment may be avoided just because of its novelty or because the animal does not recognize its value. Objects put in an animal's cage with the intention of 'enriching' the environment may only become attractive several days after the animals have become used to them.[14] But looking at the behaviour of animals when they have experienced something over and over again gets round that problem and allows us to see what the animal really wants when it knows what it is choosing.

Using this approach, we can even ask sheep how they view having their fleeces sheared and ask cows about their attitude to being shouted at or having their tails twisted. All we have to do is to repeatedly expose animals to something and then see how they react the next time around, and the next time and the next time. Do they behave as though they have had an experience they want to avoid repeating or do they behave as though they want more of it and more quickly? Do they go back for more? What is needed is some way of measuring the change in behaviour as a result of what they have experienced previously. An easy way is to set up a runway with the 'something' in question at one end. The time the animal takes to move down the runway on the first occasion is measured. It is allowed to interact with whatever it is at the far end for a certain amount of time and then it is removed and put at the start of the runway and timed again. This is done over and over again until a picture emerges. Is the animal taking more and more time to run down the runway, suggesting that it does not find the experience at the other end particularly attractive or rewarding? Or is it running faster and faster with every trial, suggesting that it wants to repeat its previous experience as quickly as possible and get to what it wants? Using change in eagerness to repeat what happened on previous occasions can be used to rank how much animals want different things, such as different kinds of enrichments—nest material, places to hide, access to companions, and so on.[15] It can also be used to reveal what they least want.

Jeff Rushen used this runway method to find out which aspects of the shearing process sheep most disliked.[16] By all obvious criteria, sheep find the whole process of being caught, held down, and sheared something they definitely don't want. They struggle violently and run away if they can. Rushen placed individual sheep in a runway which they ran down quite readily. When they got to the far end, one of three things happened. Some of the sheep were released and allowed to run straight back to their flock without having their fleeces removed. A second group were held for a few minutes in a machine designed to hold sheep for shearing but were not touched. The third group were held in the machine for the same length of time but were 'pretend-sheared'—that is, the clippers were rubbed backwards and forwards over their bodies but no wool was actually removed.

The three groups of sheep were subjected to their respective treatments a total of seven times and their subsequent reluctance or willingness to move down the runway was recorded. Sheep that were not handled at all continued to run down the runway without hesitation and became faster as time went on. Sheep that had been put in the machine, however, showed increasing reluctance to move down the runway. Most of them had to be pushed. The group that had been restrained and mock sheared, however, were the most reluctant to move of all and showed the biggest drop in running speed. Rushen concluded that both restraint in the machine and shearing were treatments the animals wanted to avoid but that actually being sheared was worst of all.

Rushen then applied this technique to ask the sheep what sort of restraint they wanted least. This was an important welfare question because of the possibility of restraining the sheep not with a mechanical restraint but with electro-immobilization. Electro-immobilization involves passing a low voltage electrical current through the body so that all the skeletal muscles contract and the animal is unable to move. The manufacturers of this electro-immobilization equipment claimed that this was a much more humane way of shearing sheep because they did not struggle and show other obvious behavioural signs of wanting to escape. Rushen showed, however, that the sheep strongly disliked the experience of being immobilized in this way. The sheep that had been electro-immobilized became more and more reluctant to approach the end of the runway the next time they were placed in it, even though by this time they had completely recovered from the shock and were perfectly well able to run fast if they wanted to. They were even more reluctant to move down the runway than the group that had experienced mechanical restraint and mock shearing. It appeared that the electro-immobilization, far from calming the sheep down, had prevented them from moving, but in a way that the sheep wanted to avoid in the future.

Rushen further argued that using a sheep's running speed in the runway gave a much more reliable indication of the sheep's view of what was being done to it than physiological measures such as corticosteroids or B-endorphin levels. The reduction in the sheep's running speed over time as they got more and more wise as to what was in store for them was directly proportional to the amount of current applied during the immobilization process in both voltage levels and duration. The physiological measures, by contrast, showed no difference between shearing, mechanical restraint, and electro-immobilization and so would not have revealed the sheep's particular dislike of immobilization.

More recently, Rushen and his team have used a version of a runway with a choice point to ask young cattle what sort of handling they want.[17] The animals were individually put at the beginning of the runway and then came to a dividing of the ways. If they went down one arm, they would be treated in one way, such as being spoken to gently ('Viens, ma belle' and other pleasantries). If they went down the other arm, they would be slapped on the rump or shouted at ('Hey, you stupid cow! Let's go, let's go, let's go!'). The heifers were first given experience of these different treatments, always one type of treatment down one arm and a second type down the other arm. After they were experienced with what happened in each arm of the runway, they were offered a choice repeatedly. The cattle showed that they had no particular preference for being spoken to in a gentle voice over not being spoken to at all, but did choose the gentle voice over being shouted at. They didn't seem to particularly mind having their tails twisted, either; at least as judged by the fact that they chose the arm where they had their tails twisted gently for three seconds equally often to an arm where a human was present but did nothing.

All these versions of choice tests—simple choices, choices in which animals have to push weights or jump over obstacles to get what they want, and repeated choices in runways—are essentially ways of getting animals to tell us what they want by doing something that comes quite naturally to them. They move towards what they want, away from what they want to avoid, and this occurs all the time in nature. Even overcoming obstacles or pushing weights is not that different from a junglefowl pushing its way through a bamboo thicket or a fox digging to get into a chicken house. Usually in these experiments, the animals can already see what they want before they make their choice, and if they can't already see or hear or smell it directly, they can see a place or an object that is associated with what they want. In striving for their goals, then, it could be argued that they are not doing anything more remarkable than what plants do when they grow towards the light. A simple attraction or repulsion, as automatic as a magnet.

However, the choices of animals can be much more complex. Animals don't just have a fixed set of 'innate' responses, a limited repertoire they are confined to. They aren't just programmed to move towards this or away from that. A major step in the evolution of wanting (or rather, mechanisms that enable animals to get what they want) was the appearance of a quite different way of making choices. Animals evolved the ability to make a huge variety of new responses and to include in their repertoire behaviour that had never been seen before in the history of their species. They could do things and make choices that neither their parents nor their grandparents had ever done. And yet the behaviour was still under the influence of genes inherited from those parents and grandparents. The secret was that genes stopped specifying specific outcomes in the form of a limited number of rigidly fixed behaviours and started specifying some much vaguer 'rewards', goals, or outcomes for animals to pursue.[18] These consisted of things like 'sweet taste' or 'warmth' as desirable or positive outcomes. Natural selection favoured animals that repeated actions that led to these positive outcomes, whatever action it was they had just done. This is why we can train animals to do tricks that are quite unlike their natural behaviour. As long as they get their 'reward', that's fine with them. Conversely, avoiding 'punishment' such as bitter taste, injury, or electric shock can be done very effectively with a huge variety of different behaviours, adapted to a wide variety of different dangers. Making rewards and punishments the basis of choice mechanisms opens up a whole range of possibilities for flexible, adaptive behaviour that are simply not open to more hidebound creatures governed by a strictly limited number of innate responses.

Edmund Rolls has argued that the evolution of choice mechanisms based on rewards and punishments rather than on innate responses was the point in evolutionary history when emotions evolved.[19] In fact, he defines emotions as states produced by rewards and punishments respectively (without implying that these states are necessarily conscious states). Animals want to obtain rewards and they want to avoid punishments, so rather than studying just what animals overtly do by way of choice, we need to study what they want by way of rewards and what they want to avoid by way of punishments. We thus have a much more general way of specifying what animals want and a much more general way of expressing how much they want it. We can compare how much animals want across a wide range of reinforcers. For example, chaffinches will learn to sit on a particular perch for the reward of simply being able to hear another chaffinch singing,[20] male Siamese fighting fish will learn to swim through a hoop for the reward of being able to see another male (and display to him).[21] Butterfly fish, which live on coral reefs and rely on smaller, cleaner fish to pick parasites off their bodies, find it rewarding just to be able to see a model of one of their cleaners. They rub up against it and if it is taken away, they will learn to operate a piece of equipment to get it back again.[22] The point about all these examples is that the animals are not just performing innate behaviours, but learning to do completely arbitrary, peculiar behaviours, operating computers, swimming through hoops, flicking switches—anything to get what they want.

W. M. S. Russell and R. L. Burch,[23] whose '3Rs' (Reduction, Replacement, and Refinement) are now widely adopted as the basis for doing experiments on animals, were some of the first to realize the importance of rewards and punishments for animal welfare. They argued that one of the best criteria for deciding whether a given set of conditions causes animals to suffer is whether those conditions could be shown to act as negative reinforcers or punishments. They saw a close link between poor welfare and negative reinforcement. On their view, animals would learn to avoid situations in which they suffered and, in turn, suffering could be recognized by whether the animals chose to avoid them. The idea of using positive and negative reinforcement as a way of finding out what animals want and what they don't want is therefore deeply embedded in the history of animal welfare.[24] It provides a very convenient way of getting them to give a detailed ranking of completely different outcomes.

For example, on most commercial pig farms, sows are not given nest material before they give birth, even though in nature pigs build large nests, and, if given straw, even modern breeds will build nests. So how important is it for pigs to be able to build nests in comparison to, say, how much they want food? We could, of course, try to answer this question by offering them a straight choice between a pile of nest material and a pile of food, but we can get a much more quantitative assessment of how important nest material is to a sow by seeing how highly she rates food and nesting material as reinforcers. Pregnant sows can easily be trained to press one button with their noses to get food and a different button to get some straw for building a nest. Then the food and nesting material can be made more difficult to get by arranging it so that a sow cannot have either food or nesting material unless she presses the button not once, but twice, four times, sixteen times, or even more. She now has to work for what she wants. She can 'buy' nesting material or she can 'buy' food and we can see exactly how she rates the two. Normally, pigs give high priority to food, but pregnant sows value nesting material so highly that, in the few days before they give birth, they will work hardest of all for nesting material, even harder than they will work for food.[25] Their motivation to build a nest at this time in their lives is so strong that they will pay a higher price for something to build a nest with than even for food and we can say precisely how much higher that price is.

Once we know what animals want and what they don't want, and have established how strong their various preferences and aversions are, this opens up a whole new range of possibilities for interpreting the rest of their behaviour. We don't have to keep giving them preference tests or making them peck keys or press buttons all the time. This is just as well because preference and operant conditioning tests are often difficult to carry out in practice. We can simply use the behaviour that the animals themselves show in the presence of, or even in anticipation of, these rewards and punishments. For example, rats that have learnt to associate different signals with future events that they find rewarding (being transferred to an enriched cage or a sexual encounter) show different 'anticipatory' behaviour from signals indicating that things they do not like are about to happen (forced swimming, transfer to a standard cage).[26] When they hear the signal, they start moving around more and keep changing behaviour more. So just by looking at the rat's behaviour, it is possible to tell whether the rat is anticipating something that it finds rewarding or something that it really doesn't want at all. The behaviour can now be correctly interpreted as that of a rat anticipating something it wants or something that it does not want because of the previous work that established which behaviour was associated with pleasurable events and which with aversive ones. Knowing what the rats find positively and negatively reinforcing is like a code book. It enables us to crack the code of rat behaviour and interpret what they do in terms of what they want and do not want to happen, even before it happens.

Of course different species have different code books. The amount of eye white a cow is showing is an indication of how frustrated she is.[27] The squeals of piglets are different depending on whether they are hungry or cold, and so on.[28] To interpret the behaviour of different species correctly, it is necessary to do a lot of background work to show which behaviour is associated with situations the animals find positive and which with situations they find negative. And it is also necessary to distinguish the behaviour they show in anticipation of those positive and negative situations (what they want and what they don't want before they get it) from behaviour they show when they are in those situations (what they like and dislike). Not everyone who wants something very much necessarily likes it when they get it.

Once we have a code book for each species, we can begin to understand their behaviour in more detail, because wanting (or not wanting) different things may be expressed in different ways. For example, behaviour associated with wanting to escape (fear) may be different from the behaviour of wanting something and being unable to get it (frustration), and different again from not having something that is wanted (deprivation). Anticipating food may be expressed differently from anticipating sex and so on.

We are just beginning to understand the ways in which animals show us, by their behaviour, what they want and what they do not want, even when no actual preference or runway or operant conditioning tests are being performed at the time. These are tedious and time-consuming. It would be almost impossible to keep offering animals choices or setting up operant conditioning tests on farms or in zoos, laboratories, or even in people's homes to see what their pets wanted all the time. Fortunately, it is not necessary to keep preference testing regularly to know what animals want. By doing preference tests under controlled conditions as the basis for writing code books for each species, we are liberated from having to keep doing them on a regular basis. Squeals, grunts, and other sounds can be documented as what a species sounds like in a variety of situations that it shows, by what it chooses, are positively and negatively reinforcing to it.[29] Body postures and specific behaviours can be classified in the same way as indicators of whether or not the animal has what it wants. Human beings who are willing to take the trouble to observe animals carefully are surprisingly good at interpreting the body postures of animals.[30] We still have a long way to go, but by using what the animals themselves want, we give the observations a solid base. We have a Rosetta Stone; a way of interpreting what we see.[31]

So, there are many ways in which we can find out what animals want. We can find out how much they want them in relation to the 'cost' they will pay for them, both in terms of arbitrary costs we humans might impose on them and in relation to the other commodities they might forgo in the real world. Wild animals have evolved by natural selection to make the most difficult of choices—to balance feeding against fleeing from the predator, protection of self against saving offspring, breathing versus sex—and we can tap into these built-in mechanisms for survival to understand what it is that animals want. Wanting and wanting the right things at the right time are the keys to survival in the wild. And when we bring animals into our environments and give them less space but more food than they have been adapted to in the wild, more freedom from predators and disease but much less freedom to move around, they bring with them needs and wants evolved over millions of years. To keep them alive, we have to satisfy their basic physical needs, but why should we care about their wants? Why shouldn't we just have a 'one-question' definition of animal welfare and simply satisfy their needs so that they are kept in good physical health? Why should we bother at all about what animals want?

There are two reasons for including 'what the animal wants' as an essential part of animal welfare. The first is the close connection between the two questions. For purely practical and self-interested reasons we regard animal health as important to human health, so for exactly the same selfish reasons, what the animals themselves want also becomes important to us. When animals do not have what they want, they show signs of frustration (being unable to obtain what they want), boredom (lack of stimulation), fear (wanting to escape), or deprivation (wanting something that isn't there). Many of these behaviours are quite 'natural' in the sense that they are seen in wild animals. In wild animals, however, they are short-lived means to an end. An animal trapped in a thicket will be fearful, frustrated that it cannot escape, and will, as a result of wanting to escape, move backwards and forwards until, in all likelihood, it finds a way out. The fear, the frustration, and the wanting to escape are all part of natural—and usually highly effective—escape mechanisms that successful animals have evolved for survival. The escape mechanisms are activated, they work, and then they stop. It is only when the animal is put in a cage (effectively, an impenetrable thicket that is impossible to escape from) that the fear and frustration and the wanting to escape don't work any more. They are activated and they keep being activated because the goal (escape) is never achieved. They are like an inefficient heating system of a house that is constantly activated by cold weather but never switches off because it never makes the house quite warm enough to achieve its goal. The fact that the heating never switches off and never achieves the temperature it is supposed to is itself a sign that something is wrong. The system is ineffective, inefficient, and probably costs a lot of money, not to mention making an unnecessary contribution to global warming.

With animals, too, constantly activated mechanisms associated with unmet wants are a sign that something is wrong. They might be precursors of trouble ahead—preclinical indicators or an early warning system. Even those people who see animals as insentient machines can see the sense in having a well-tuned, properly functioning machine that is more likely to stay functioning for longer than a machine that is struggling. We oil engines to prevent future wear and tear. We tune them to get the maximum use out of them. In the same way, an animal with unmet wants is inefficient and unhealthy, if for no other reason than that it is spending a great deal of energy 'trying' to get what it wants and possibly damaging itself in the process.

Similarly, what an animal wants—and particularly signs that it doesn't have what it wants for prolonged periods of time—can be an indicator that the machine is not functioning as well as it might. But including 'what the animal wants' as part of the definition of good welfare has another advantage too. It satisfies those people who think that there is more to good welfare than just physical health. 'What animals want' takes amorphous and difficult concepts such as 'mental health', 'positive affective state', and 'quality of life' and pins them down. It makes them understandable and it shows us in a clear and down to earth way what we have to find out in order to demonstrate good or bad welfare. It provides the 'more' in the phrase 'more to good welfare than just good health'. But in giving 'what animals want' such a prominent position in the definition of animal welfare, aren't we assuming that the 'wanting' must indicate that it is conscious wanting? By putting so much emphasis on what animals want haven't we come full circle and based the definition of 'animal welfare' once again on the assumption that animals are consciously aware?

The answer is no, we have not. The mystery of consciousness remains. The explanatory gap is as wide as ever and all the wanting in the world will not take us across it. Animals and plants can 'want' very effectively with never a hint of consciousness, as we can see with a tree wanting to grow in a particular direction. Preference tests, particularly those that provide evidence that animals are prepared to pay 'costs' to get what they want, are perhaps the closest we can get to what animals are feeling,[32] but they are not a magic entry into consciousness. They do not solve the hard problem for us because everything that animals do when they make choices or show preferences or even 'work' to get what they want could be done without conscious experience at all. We have seen (Chapters 4 and 5) just how much we humans do unconsciously and how powerful our unconscious minds are in making decisions and even in having emotions. What is good enough for us may well be good enough for other species.

The most that can be said is that the chasm between what we can study scientifically (behaviour and physiology) and consciousness is perhaps made a little narrower by showing that animals will work to obtain one set of circumstances and avoid others. But the gap is still there and we still have no sure way of crossing it. Even with human beings, we do not know what other people experience in the way that each one of us knows, from the inside, what we experience.

In the case of other humans, we use words to ask them what they are feeling, and use what they say as a reasonable working substitute for direct knowledge of what they are experiencing. Preference tests and their variations could be seen as the animal equivalents of asking people in words and it is tempting to say that they are as good as words, if not better. So if we are happy enough to use words as a rickety bridge across the chasm, why not use preference tests, choice, and operant conditioning to do the same for animals? This argument seems particularly compelling when we look at the evidence that animals will choose to give themselves the same drugs that we know have pain-relieving or anxiety-relieving properties in ourselves. Isn't this direct evidence for conscious experience of pain in animals? Doesn't this show that their pain is like ours, not just in the external symptoms that they show but also in what they *feel*?

For example, broiler (meat) chickens that are unable to walk without limping will learn to eat food containing carprofen, a drug known to relieve pain in humans. If given two kinds of food coloured in different ways, such as red for food with carprofen and blue for food without it, severely lame chickens will taste both but then come to prefer the red food containing the pain reliever.[33] They start eating much more of it and, most remarkably, their lameness symptoms start to disappear. They start walking better and more freely. Healthy chickens that previously had no difficulty walking show no such preference for eating food with pain-reliever, so this preference for the drug is quite specific to lame birds. This means that the chickens' behaviour—wanting the drug when they have difficulty walking and then walking more easily after they have taken it—is very similar to that of a human with backache reaching for the medicine bottle and feeling better afterwards. And it isn't just chickens that like our drugs. Sick sheep, mice, and arthritic rats do the same.[34] In fact, human pain-relieving drugs are often tested on animals precisely because their responses are so similar to ours. Mice will even give themselves drugs known to have the effect in humans of reducing anxiety,[35] suggesting similarities that go beyond just pain reduction and take us into the realm of reducing other unpleasant emotional states too.

The similarity between the behavioural responses of animals and humans to such drugs makes it tempting to assume that because the behaviour is similar, the conscious experiences must be similar too. Of course they may be, but there is no more 'must' about it than in the claim that animals 'must' consciously experience thirst before they drink or 'must' consciously experience hunger while they are searching for food. They may well do so, as we saw in Chapter 8. But there is no must about it. Animal bodies have evolved by natural selection to restore imbalances of food and water and to repair wounds and other kinds of damage. Neither food deprivation nor water deprivation, nor the symptoms of inflamed joints, are necessarily accompanied by any conscious experiences at all, although they may be. Just as our wounds heal up without any conscious intention on our part and we like certain foods without knowing why, so other animals, too, have a variety of mechanisms for repairing and restoring their bodies to proper working order.[36] Preference and choice and 'what animals want' are part of these mechanisms. They may well be accompanied by conscious experiences. But then again, they may not be. Once again, our path to finding the answer one way or the other is blocked by the implacable, infuriating obstacle known as the hard problem.

All we can say with any certainty is that living organisms have the most remarkable capacity for staying alive, repairing themselves, finding food and water, escaping danger, and even anticipating danger. But exactly where consciousness crept in, no one knows. I hope this book has convinced you that being a consciousness sceptic is not to be anti-animal. It might seem to help animals by claiming that the 'problem' of consciousness has been solved and that all we need now is a bit of anthropomorphism and all will be fine. But, in reality, all this does is to make animal welfare look unrigorous and unscientific.

No, consciousness-scepticism helps animals. Animal welfare needs new arguments if it is to hold its own against the competing claims now being made on the world's attention. First, it needs the best scientific evidence available, not wishful thinking or anthropomorphism. Second, it needs to be linked to concerns that even people who currently care little or nothing about non-human animals cannot ignore—issues such as their own health, welfare, and quality of life. Let us now see what that means when we step back and take a look at the welfare of animals in the wider context of all these other issues that people are also concerned about.

10

ANIMAL WELFARE
FOR A SMALL PLANET

Saying that animal welfare needs new arguments is not to say that the arguments that are currently used to support the welfare of non-human animals are invalid. But they do need boosting with additional evidence and additional arguments aimed at those who have so far not seen the point of animal welfare at all.

The aim of this book has been to show what these arguments might be and how the future case for animal welfare might best be made. I have questioned what has now become the new orthodoxy about animal welfare—that anthropomorphism is all we need and that scientists who point out the problems of attributing conscious experiences to animals are out of date and, worse, are holding back progress in animal welfare.[1, 2] Once, it was scientists who dared to discuss animal consciousness and who were criticized as being unscientific. Now there is a new set of taboos and it is those who point out how difficult consciousness is to study who are derided and ridiculed.[1]

I have argued that animal welfare would benefit from a much more open-minded and critical attitude to animal consciousness. The case for animal welfare does not stand or fall by the claim that animals are conscious like us, so we should not worry that being critical of the evidence will damage the cause of animal welfare. On the contrary, by acknowledging what a real problem consciousness still is, we clear the way for a truly scientific study of animal welfare, one not based on anthropomorphism. Rather than claiming to have solved the problem of consciousness, it is much, much better *for animals* if we remain sceptical and agnostic. Militantly agnostic if necessary, because this keeps alive the possibility that a large number of species have some sort of conscious experiences, rather than ruling them out because they do not fit a particular theory of what consciousness is or appear to fall foul of a 'killjoy' explanation.[3] For all we know, many animals, not just the clever ones and not just the overtly emotional ones, also have conscious experiences. A consciousness-sceptic would not even rule it out for invertebrates such as crabs and prawns.[4]

So we shouldn't just be facing up to the hard problem of consciousness. We should be standing up for it and shouting for it. Animal welfare benefits from acknowledging how hard that problem is and derives much greater credibility from evidence that does not rely on either solving the hard problem or pretending it isn't that big a deal anyway. We have only just begun to scratch the surface of the possibilities of what those other arguments might be, largely because of too much compartmentalizing the way people think and the way research is funded. 'Animal welfare' has been seen as somehow separate from other things, an expensive extra to be added on, rather than as an integral part of keeping animals healthy and even of preventing disease altogether.[5]

The way forward must be to find solutions that both benefit humans *and* ensure the welfare of animals. This will need research and all the resources that animal welfare science can muster, but there is evidence that it can be done. Temple Grandin,[6] for example, has shown that small changes to the way animals are treated in slaughter houses can result in greater efficiency, lower costs, *and* higher welfare. The lesson we learn from Grandin's research is that we may have to put in considerable effort to find solutions that work in practice. They may not be obvious at first. They may not even work well the first time something new is tried. They may need extensive development to make sure that they do work, and that they are affordable and practical. The point is that somehow, in searching for those solutions, we make sure that animal welfare is one of the many priorities that need to be taken into account. All the rest *and* animal welfare. Not *or* animal welfare.[7] But animal welfare up there with the other priorities—human health, human well-being, sustainable food production, and protecting the planet.[8]

In many cases, we simply don't know whether animal welfare is the ally or the antagonist of other factors. There is a lot of work to do in finding out and then in finding solutions that fulfil all our goals. Many people want to see animals living free-range for example, but they also want their food to be healthy and free of contamination by disease organisms. So for good welfare the animals would be outside, but to protect people from disease, they would be inside behind a biosecurity barrier. This would suggest that food safety and animal welfare are in direct conflict, but of course the answer does not come from armchair speculation as to whether or not they are. It has to come from finding out what actually happens. Are animals more prone to disease if they are kept outside because they are in contact with more disease organisms, or are they less prone to disease because their immune systems are better and they are better at fighting off disease? Is animal welfare the helper of disease resistance or a cost to be paid in terms of greater incidence of disease? Is it a benefit or a cost, and if it is a cost, how much of a cost is it and are people willing to pay that cost?

If there is one lesson you take away from this book, I hope it would be this: we don't yet know the answers to such questions and we desperately need more science to help us to find them. We need facts and we need proper scientific evaluation of such facts. That means good science, clear definitions, wide-scale measurement, and proper analyses. We need studies that evaluate not just animal welfare (in isolation) or food safety (in isolation) or environmental impact (in isolation) or production costs (in isolation), but give us the full picture, across the board, of what happens with different systems.

Obviously some people will hope the results come out in favour of the free-range system, while others will be keeping their fingers crossed that the intensive system is, on balance, better all round. The answers may be different in different situations, in different countries, and with different species.[9] And even when we have the first answers, we may be able to modify what doesn't work and turn it into what does provide the best overall balance. That means that when sustainable intensification is discussed at the world's conference tables, there may not need to be a separate seat labelled 'animal welfare', with representatives speaking from just this one point of view. It may be that by far the most effective way to get real improvements in animal welfare implemented across a world, much of which does not care about animal welfare, is to speak through delegates, whose primary concern is with human welfare and let them do the talking on behalf of animals. To be convinced to do this, they will need hard evidence about how animal welfare contributes to the matters they are most concerned about (food production, human health, environment). But they will also need tried and tested practical solutions that are cost-effective and acceptable to the people who are going to have to implement them.

If animal welfare science is to live up to its name, it needs to be able to fulfil this role of evidence-provider and evidence-based solution provider much more effectively than it does at present. It needs to have more data and more information about what works and what doesn't. We need a revolution in the way data is collected and even more in the way it is analysed and dealt with. The case for animal welfare needs to be made with much better information than is currently available, so that it is less reliant on what people think is best for animals and more reliant on what actually is best for animals. We need more science not less science, but that science needs to be more multidisciplinary and able to show the connections (or, it may turn out, the lack of them) between animal welfare, human health, food production, and other concerns.

This change in animal welfare from a single isolated cause where change is driven largely by pressure groups to a science-based animal welfare integrated with other concerns needs three things. First, it needs clear objectives or outcomes. It needs a definition of 'good welfare' that everyone can buy into and that is easy to measure. I have suggested that 'good health and animals having what they want' is a workable definition, on which the many different measures of welfare that have now been suggested can be hung. All measures relate to the proper functioning of the organism, no more mysterious (although perhaps more complex) in animals than in plants. Good health, as indicated both by current symptoms and by future intimations that all may not be well, can be used here. A word of warning. Good welfare is a property of individuals, not groups or farms or 'systems'. So in using health and proper functioning in this way, it is important to specify that it is the health and well-being of the animal that is important. A farm might be productive and have overall reasonable health levels, but individual animals might have to be culled as a result of being injured or diseased. This is not good individual welfare. The divergence between what is good for individual animal welfare and what is best for the group is least when the intrinsic value of each animal is high. With racehorses, the health and well-being of each individual horse is directly in line with the interests of the stables. But with broiler chickens, the cash value of each individual animal is low. This means that a loss of health or even death of one chicken makes very little difference to the cash value of the flock. We need to keep our eyes on the individual.

Secondly, these welfare measurements need to be made on a wide scale in the situations in which they apply. For example, although preliminary measurements on farm animals in small pens are a valuable first step, the results of any small-scale studies need to be scaled up and tested on real farms. Real farms have problems with water pipes freezing in winter or problems with getting good staff, which may 'throw' a well-designed system so that although it works well on a small supervised scale, it results in poor welfare if anything goes wrong out there in the real world. We have at present far too little data of the sort that is needed for the wide-scale evaluation of welfare. Some of this data is already available and collected by commercial companies but too commercially sensitive to be released. We need more non-invasive and automated ways of assessing welfare so we can constantly assess what is going wrong and what is going right.

Thirdly, we need good science to evaluate the data. Human medical science has now established ways, such as the systematic Cochrane-type reviews, of taking all the scientific data available and coming up with proper evaluations that go beyond cherry-picking, beyond personal prejudice, and towards a review of the best available evidence.[10] A robust and evidence-based evaluation of the best available data, not just on animal welfare but on how welfare relates to human health, human food production, and environmental issues, would be most likely to be noticed by policy-makers. Truly science-based animal welfare is going to be the most important ally of animal welfare and the most likely way to make the voices of non-humans heard.

In rethinking animals and our attitudes to animal welfare, we need to be prepared for rearrangements of the mind, which are never comfortable but essential for progress. There is nothing to stop anyone believing that animals have conscious experiences. Indeed, this will probably continue to be the main driver for many people, the reason why they think animal welfare is important. But the argument of this book has been to point out that this may not be enough to convince some people, particularly in a world increasingly concerned with human issues and human crises.

Animal welfare is in danger of being drowned out in the calls for increased food production, the desire of increasing numbers of people to have a higher standard of living, and the steps being taken to mitigate the effects of climate change. With more people and rising living standards, there is more and more demand for animal products. We will all probably have to get used to using less of these products simply because they will become more expensive. But campaigns to persuade people to eat less will be counteracted by the greater numbers of people wanting to eat meat, eggs, and milk. In the longer term, we may find acceptable foods based on non-animal products. We may all end up eating algae. But in the short term, by which I mean now and the next few years, animal products are in demand, land is in demand, and intensification is the watchword. With rising human numbers and rising human expectations, the welfare of non-human animals is in danger of being ignored or trampled on or lost. That is why we have to rethink our attitudes to animals. Those attitudes and the arguments we use will be crucial to whether or not other people are likely to be convinced that animal welfare is worth taking seriously.

We are up against culture, religion, and a widespread and deep-seated conviction among many people that humans are much more important than any animal. The case for animals based on their capacity to suffer and experience has been made with eloquence by people such as John Webster, Mary Midgley, Peter Singer, Richard Ryder, and others.[11] There is no question that the welfare of animals has improved immeasurably over the past fifty years in all areas—farming, zoos, laboratories, pest control—although rather more in some areas than others, and, in many people's eyes, not enough. But we are now at a new point in the history of animal welfare, where emotional appeals have gone so far but where they may not be enough to go the next step and ensure that animal welfare retains and improves its position as a societal goal. The signs are not good. International and government reports on the future of the world do not give animal welfare a priority. Perhaps the title of Bernie Rollin's book, *The Unheeded Cry*,[12] should be taken not just as a call for action but as a description of the current situation. If the cry hasn't been heeded, then that is a sign that new arguments are needed.

The current fashion for science-bashing and ridiculing scientists for pointing out how hard it is to study consciousness is not helping the development of a scientifically based study of animal welfare. Nor is the lack of evidence-based studies linking animal welfare to wider concerns such as human health and productive ways of raising animals. Animal welfare needs more science, not less. It also needs to be seen not as an isolated fringe interest, based on vague ideas of what might be good for animals, but as linked to a much wider range of concerns that everyone can see as affecting their own future. The earth is turning out to be a much smaller planet than it once seemed, but we humans are not its only inhabitants. We need to rethink our view of the millions of non-human animals that also live here, not just in regard to what they are in themselves, but also in how our own futures are inseparably bound up with theirs.

NOTES AND REFERENCES

CHAPTER 1

1. H. Steinfeld, P. Gerber, T. Wassener, V. Castel, M. Rosales, and C. de Haan (2006) *Livestock's Long Shadow: Environmental Issues and Options*. Food and Agricultural Organization of the United Nations, Rome.

2. Foresight (2011) *The Future of Food and Farming: Challenges and Choices for Global Sustainability*. Final project report. The Government Office for Science, London. This report lists the main drivers for change as:

    A. Balancing future demand and supply sustainably—to ensure that food supplies are affordable.
    B. Ensuring that there is adequate stability in food supplies—and protecting the most vulnerable from the volatility that does occur.
    C. Achieving global access to food and ending hunger.
    D. Managing the contribution of the food system to the mitigation of climate change.
    E. Maintaining biodiversity and ecosystem services while feeding the world.

3. H. J. Godfray, J. R. Beddington, I. R. Crute, L. Haddad, D. Lawrence, J. F. Muir, J. Pretty, S. Robinson, S. M. Thomas, and C. Toulmin (2010) Food security: the challenge of feeding 9 billion people. *Science* 327: 812–17.

4. Peter Singer (1976) *Animal Liberation: A New Ethics for Our Treatment of Animals*. Jonathan Cape, London. The first chapter of this landmark book is entitled 'All animals are equal or why supporters for blacks and women should support animal liberation too'. Singer acknowledges that he owes the term 'speciesism' itself to Richard Ryder (1975) *Victims of Science*. Davis-Poynter, London.

5. Two contrasting views: M. Midgley (1983) *Animals and Why They Matter: A Journey Around the Species Barrier*. Pelican Books, Harmondsworth; T. R. Machan (2004) *Putting Humans First: Why We are Nature's Favorite*. Rowman and Littlefield, Lanham, MD.

6. J. Houghton (2009) *Global Warming: The Complete Briefing*, 4th edn. Cambridge University Press.

7. B. E. Rollin (2007) Cultural variation, animal welfare and telos. *Animal Welfare* 16: 129–33; A. Miura and J. W. S. Bradshaw (2002) Childhood experience and attitude towards animal issues: A comparison of young adults in Japan and the UK. *Animal Welfare* 11: 437–48.

8. R. Harrison (1964) *Animal Machines: The New Factory Farming Industry*, foreword by Rachel Carson. Vincent Stuart, London.

9. D. R. Griffin (1976) *The Question of Animal Awareness: Evolutionary Continuity of Mental Experience*. Rockefeller University Press, New York.

10. M. Bekoff (2007) *The Emotional Lives of Animals*. New World Library, Novato, CA.

CHAPTER 2

1. Enrichment: R. R. Swaisgood et al. (2001) A quantitative assessment of the efficacy of an environmental enrichment programme for giant pandas. *Animal Behaviour* 61: 447–57; R. C. Newberry (1995) Environmental enrichment: increasing the biological relevance of captive environments. *Applied Animal Behaviour Science* 44: 229–43.

2. C. M. Sherwin, G. J. Richards, and C. J. Nicol (2010) Comparison of the welfare of layer hens in four housing systems in the UK. *British Poultry Science* 51: 488–99.

3. D. Chalmers (1995) Facing up to the problem of consciousness. *Journal of Consciousness Studies* 3: 200–19. Reprinted as: 'The hard problem of consciousness' and 'Naturalistic dualism'. In *The Blackwell Companion to Consciousness* ed. M. Velmans and S. Schneider (2007) Blackwell, Oxford, 225–35 and 359–68 respectively.

4. T. Nagel (1974) What is it like to be a bat? *Philosophical Review* 83: 435–50; D. C. Dennett (1996) *Kinds of Minds: Towards an Understanding of Consciousness*. Weidenfeld and Nicolson, London; S. Budiansky (1998) *If a Lion Could Talk: Animal Intelligence and the Evolution of Consciousness*. The Free Press, New York.

5. J. Balcombe (2006) *Pleasurable Kingdom: Animals and the Nature of Feeling Good*. Macmillan, London; M. Bekoff (2007) *The Emotional Lives of Animals*. New World Library, Novato, CA.

CHAPTER 3

1. J. S. Kennedy (1992) *The New Anthropomorphism*. Cambridge University Press.

2. M. Bekoff (2007) *The Emotional Lives of Animals*. New World Library, Novato, CA.

3. D. L. Cheney and R. M. Seyfarth (1990) *How Monkeys See the World: Inside the World of Another Species*. University of Chicago Press, Chicago. p. 303.

4. F. B. M. de Waal (1996) *Good Natured: the Origins of Right and Wrong on Human and Other Animals*. Harvard University Press.

5. R. I. M. Dunbar (1984) *Reproductive Decisions: Economic Analysis of Gelada Baboon Social Strategies*. Princeton University Press, Princeton, NJ.

6. J. S. Kennedy, *The New Anthropomorphism*.

7. M. Bekoff (2007) *The emotional Lives of Animals*. New World Library, Novato, CA, p. 120.

8. Ibid. p. xviii.

9. M. Bekoff (2007) Do animals have emotions? *New Scientist*, 23 May. http://www.newscientist.com/article/mg19426051.300-do-animals-have-emotions.html (accessed 10 November 2011).

10. N. Tinbergen (1951) *The Study of Instinct*. Oxford University Press, Oxford, p. 4.

11. Ibid., p. 5.

12. J. B. Watson (1929) *Psychology from the Standpoint of a Behaviorist.* Lippincott, Phil.

13. J. B. Watson (1913) Psychology as the behaviorist views it. *Psychological Review* 20: 158–77.

14. B. F. Skinner (1938) *The Behavior of Animals in Experimental Analysis.* Appleton-Century-Crofts, New York.

15. J. S. Kennedy, *The New Anthropomorphism* p. 2.

16. D. R. Griffin (1976) *The Question of Animal Awareness: Evolutionary Continuity of Animal Awareness.* Rockefeller University Press, New York.

17. M. S. Dawkins (1980) *Animal Suffering: the Science of Animal Welfare.* Chapman and Hall, London.

18. G. Burghardt (2004) Ground rules for dealing with anthropomorphism. *Nature* 430: 15.

19. M. Bekoff (2007) *The Emotional Lives of Animals.* New World Library, Novato, CA, p. xx.

20. Ibid. p.123.

21. C. D. L. Wynne (2004) The perils of anthropomorphism. *Nature* 428: 606.

22. J. S. Kennedy, *The New Anthropomorphism.*

23. J. Goodall (1990) *Through a Window.* Houghton-Mifflin, Boston, p. 8.

24. A. Whiten and R. Byrne (1988) Tactical deception in primates. *Behavioural and Brain Sciences* 11: 233–44.

25. M. S. Dawkins (2007) *Observing Animal Behaviour.* Oxford University Press, Oxford.

26. O. Pfungst (1911) *Clever Hans, the Horse of Von Osten.* Holt, Rinehart and Winston, New York. (Reprinted 1965 in English translation by C. L. Rahn.)

27. This is discussed in more detail in the next chapter.

28. P. Aggarwal and A. L. McGill (2007) Is that car smiling at me? Schema congruity as a basis for evaluating anthropomorphized products. *Journal of Consumer Research* 34: 468–79.

29. D. Proudfoot (2011) Anthropomorphism and AI: Turing's much misunderstood imitation game. *Artificial Intelligence* 175: 950–57. Many people who work with computers have a tendency to anthropomorphize and ascribe the machines with human-like qualities.

30. http://support.sony-europe.com/aibo (accessed 10 November 2011).

31. M. Bekoff (2007) *The Emotional Lives of Animals,* p. 33: 'Animals display flexibility in their behaviour patterns and this shows that they are conscious and passionate and not merely "programmed" by genetic instinct to do "this" in one situation and "that" in another situation.'

32. Ibid.

33. J. Balcombe (2006) *Pleasurable Kingdom,* p.55, 'Machines are not flexible'.

34. M. Bekoff (2007) *The Emotional Lives of Animals.*

35. D. R. Griffin (1992) *Animal Minds.* Rockefeller University Press, New York.

36. B. R. Duffy (2003) Anthropomorphism and the social robot. *Robotics and Autonomous Systems* 42: 177–90.

37. J. W. Odell and J. Dickson (1984) Eliza as a therapeutic tool. *Journal of Clinical Psychology* 40: 942–5.

38. J. S. Kennedy, *The New Anthropomorphism.*

39. C. Wynne (2004) The Perils of Anthropomorphism. *Nature* 428: 606.

40. Ibid.

41. F. B. M. de Waal (2000) Primates—a natural history of conflict resolution. *Science* 289: 586–90.

42. S. Blackmore (2003) *Consciousness: An Introduction.* Hodder & Stoughton, London.

CHAPTER 4

1. D. Chalmers (1995) Facing up to the problem of consciousness. *Journal of Consciousness Studies* 3: 200–19.

2. N. Block (1998) How can we find the neural correlates of consciousness? *Trends in Neuroscience* 19: 456–9.

3. D. Chalmers, Facing up to the problem of consciousness. *Journal of Consciousness Studies* 5: 200–19.

4. J. Levine (2001) *The Purple Haze: The Puzzle of Consciousness.* Oxford University Press, New York.

5. Please see Chapter 3.

6. D. R. Griffin (1992) *Animal Minds.* University of Chicago Press. Griffin discusses beavers building dams on pp. 87–100 and his own first-hand observations of lions hunting cooperatively on p. 64.

7. J. Call and M. Tomasello (2008) Does the chimpanzee have a theory of mind? 30 years later. *Trends in Cognitive Science* 12: 187–92. Theory of Mind and Neural Correlates Consciousness.

8. D. C. Dennett (1983) Intentional systems in cognitive ethology: the 'Panglossian paradigm' defended. *Behavioral and Brain Sciences* 6: 343–96.

9. S. J. Shettleworth (2010) Clever animals and killjoy explanations in comparative psychology. *Trends in Cognitive Sciences* 14: 477–81.

10. C. Heyes (2008) Beast machines? Questions of animal consciousness. In *Frontiers of Consciousness*, ed. L. Weiskrantz and M. Davies. Oxford University Press, Oxford, pp. 259–74; C. Heyes (1998) Theory of mind in non-human primates. *Behavioral and Brain Sciences* 21: 101.

11. S. Coussi-Korbel (1994) Learning to outwit a competitor in mangabeys (*Cercocerbus torquatus torquatus*). *Journal of Comparative Psychology* 108: 169–71.

12. The many and complex ways that genes have of producing their effects are described in M. Ridley (2003) *Nature via Nurture: Genes, Experience and What Makes us Human.* Fourth Estate, London.

13. D. M. Rosenthal (1993) Thinking that one thinks. In *Consciousness*, ed. M. Davies and G.W. Humphreys, 197–223. Blackwell, Oxford.

14. P. Carruthers (2004) Suffering without subjectivity. *Philosophical Studies* 121: 99–125; P. Carruthers (2005) Why the question of consciousness may not matter very much. *Philosophical Psychology* 18: 83–102.

15. R. R. Hampton (2001) Rhesus monkeys know when they remember. *Proceedings of the National Academy of Sciences* 98: 5359–62.

16. D. Chalmers (1995) The puzzle of conscious experience. *Scientific American*, December: 62–8.

17. I. J. H. Duncan (1993) Welfare is what animals feel. *Journal of Agricultural and Environmental Ethics* (Suppl.) 6: 8–14; D. M. Broom (1998) Welfare, stress and the evolution of feelings. *Advances in the Study of Behavior* 27: 317–403; M. S. Dawkins (2000) Who needs consciousness? *Animal Welfare* 10: 519–29.

18. For a discussion of animal pain see P. Bateson (1991) Assessment of pain in animals. *Animal Behaviour* 42: 827–39.

19. www.thefactsofpainlesspeople.com (accessed 10 November 2011).

20. People unable to feel pain may have no obvious anatomical defects and their condition may come from a biochemical defect such as the passage of sodium ions: J. J. Cox et al. (2006) An SCN channelopathy causes congenital inability to experience pain: sodium channels. *Nature* 444: 894–8.

CHAPTER 5

1. Oxygen in the blood is carried by a molecule called haemoglobin, which has different magnetic properties depending on whether it is oxygenated (carrying a lot of oxygen) or has become deoxygenated through giving up its oxygen to cells that need it. This difference in magnetic properties of oxygenated and deoxygenated haemoglobin in different parts of the brain is picked up during fMRI to give a measure of blood flow which is, in turn, a measure of nerve cell activity.

2. C. Koch (2004) *The Quest for Consciousness*. Roberts and Company, Colorado.

3. M. Bekoff and P. W. Sherman (2004) Reflections on animal selves. *Trends in Ecology and Evolution* 19: 176–80.

4. P. Stoerig (2007) Hunting the ghost: towards a neuroscience of consciousness. In *The Cambridge Handbook of Consciousness*, ed. P. D. Zelazo, M. Moscovitch, and E. Thompson. Cambridge University Press, pp. 707–30.

5. E. Morsella, S. C. Krieger, and J. A. Bargh (2010) Minimal neuroanatomy for a conscious brain: homing in on the networks constituting consciousness. *Neural Networks* 23: 14–15; B. Merker (2007) Consciousness without a cerebral cortex: a challenge for neuroscience and medicine. *Behavioural and Brain Sciences* 30: 63+; S. Zeki (2003) The disunity of consciousness. *Trends in Cognitive Sciences* 7: 214–18.

6. J. Kulli and C. Koch (1991) Does anaesthesia cause loss of consciousness? *Trends in Neuroscience* 14: 6–10.

7. M. L. Phillips et al. (2004) Differential neural responses to overt and covert presentations of facial expressions of fear and disgust. *NeuroImage* 21: 1484–96.

8. S. Dehaene, L. Naccache, L. Cohen et al. (2001) Cerebral mechanisms of word masking and unconscious repetition priming. *Nature Neuroscience* 4: 752–8.

9. G. Rees and C. Frith (2007) Methodologies for identifying the neural correlates of consciousness. In *The Blackwell Companion to Consciousness*, ed. M. Velmans and S. Schneider. Blackwell, Oxford, pp. 553–66.

10. E. T. Rolls (2008) *Memory, Attention and Decision-making*. Oxford University Press, Oxford.

11. E. T. Rolls, M. K. Kringelbach, and I. E. T. De Araujo (2003) Different representations of pleasant and unpleasant odours in the human brain. *European Journal of Neuroscience* 18: 695–703.

12. E. T. Rolls and G. Deco (2010) *The Noisy Brain: Stochastic Dynamics as a Principle of Brain Function*. Oxford University Press, Oxford.

13. H. R. Beresford (1999) Brain death. *Neurologic Clinics* 17: 295.

14. C. Umiltà (2007) Consciousness and control of action. In *The Cambridge Handbook of Consciousness*, ed. P. D. Zelazo, M. Moscovitch, and E. Thompson. Cambridge University Press, Cambridge, pp. 327–51.

15. Different routes to action: E. T. Rolls (2005) *Emotion Explained*. Oxford University Press, Oxford, pp. 411–18.

16. L. Weiskrantz (1997) *Consciousness Lost and Found*. Oxford University Press, Oxford.

17. A. M. Colman, B. D. Fulford, D. Omtzigt, and A. al-Nowaihi (2010) Learning to cooperate without awareness in multiplayer miniomal social situations. *Cognitive Psychology* 61: 201–27.

18. R. S. Barton and P. H. Harvey (2000) Mosaic evolution of brain structure in mammals. *Nature* 405: 1055–8; T. W. Deacon (1992) The human brain. In *The Cambridge Encylopedia of Human Evolution*, ed. S. Jones, R. Martin, and D. Pilbeam. Cambridge University Press, Cambridge, pp. 115–23.

19. A. K. Seth, B. J. Baars, and D. B. Edeleman (2005) Criteria for consciousness in humans and other mammals. *Consciousness and Cognition* 14: 119–39; D. B. Edelman, B. J. Baars, and A. K. Seth (2005) Identifying hallmarks of consciousness in non-mammalian species. *Consciousness and Cognition* 14: 169–87.

20. B. Bermond (2001) A neurophysical and evolutionary approach to animal consciousness and animal suffering. *Animal Welfare* 10 Supplement: S47–S62; B. Baars (2001) There are no known differences in fundamental brain mechanisms of sensory consciousness between humans and other mammals. *Animal Welfare* 10 Supplement: S31–S40. For a different view: E. M. Macphail (1998) *The Evolution of Consciousness*. Oxford University Press, Oxford.

CHAPTER 6

1. J. Bentham [1789] (1961) Introduction to the Principles of Morals and Legislation. In *The Utilitarians*. (2001) ed G. Sher, Hackett Publishing Co., Indianapolis.

2. C. Darwin [1872] (1965) *The Expression of the Emotions in Man and Animals.* University of Chicago Press, Chicago.

3. K. Oatley and J. M. Jenkins (1996) *Understanding Emotions.* Blackwell, Oxford.

4. J. Panksepp (1998) *Affective Neuroscience.* Oxford University Press, New York.

5. A. Boissy et al. (2007) Assessment of positive emotions in animals to improve their welfare. *Physiology and Behavior* 92: 375–97.

6. Ibid. p. 376 and p. 24.

   D. Fraser (2008) *Understanding Animal Welfare: the Science in its Cultural Context.* Universities Federation for Animal Welfare, Wiley-Blackwell, Chichester.

   D. M. Broom (1998) Welfare, stress and the evolution of feelings. *Advances in the Study of Behavior* 27: 371–403.

   V. Braithwaite (2010) *Do Fish Feel Pain?* Oxford University Press, Oxford.

7. M. Cabanac (1992) Pleasure: the common currency. *Journal of Theoretical Biology* 155: 173–200; M. Cabanac (1971) Physiological role of pleasure. *Science* 173: 1103–7.

8. M. Cabanac, A. J. Cabanac, and A. Parent (2009) The emergence of consciousness in phylogeny. *Behavioural Brain Research* 198: 267–72.

9. M. Cabanac (2009) Do birds experience sensory pleasure? *Evolutionary Psychology* 7: 40–7.

10. J. Panksepp (2005) Affective consciousness: core emotional feelings in animals and humans. *Consciousness and Cognition* 14: 30–80.

    J. Panksepp (2006) The neurobiology of positive emotions. *Neuroscience Biobehavior Reviews* 30: 173–87.

11. D. Denton, R. Shade, F. Zamarippa et al. (1999) Neuroimaging of genesis and satiation of thirst and an interceptor-driven theory of origins of primary consciousness. *Proceedings of the National Academy of Sciences* 96: 5304–9.

12. J. E. LeDoux (2000) Emotion circuits in the brain. *Annual Review of Neuroscience* 23: 155–84.

J. E. LeDoux (2005) Contributions of the amygdala to emotion processing: from animal models to human behaviour. *Neuron* 48: 175–87.

13. J. T. Cacioppo, D. J. Klein, G. C. Berntson, and E. Hatfield (1993) The psychophysiology of emotion. In *Handbook of Emotions*, ed. M. Lewis and J. M. Hatfield. Guilford, New York, pp. 119–42.

14. A. Damasio (1999) *The Feeling of What Happens: Body and Emotion in the Making of Consciousness*. Harcourt Brace, New York.

15. S. T. Murphy and R. B. Zajonc (1993) Affect, cognition and awareness in affective priming with optimal and suboptimal stimulus exposure. *Journal of Personality and Social Psychology* 64: 723–9.

W. Sato and S. Aoki (2006) Right hemispheric dominance in processing of unconscious negative emotion. *Brain and Cognition* 62: 261–6.

J. S. Morris, A. Öhman, and R. J. Dolan (1998) Conscious and unconscious emotional learning in the human amygdala. *Nature* 393: 467–70.

16. M. Tamietto et al. (2009) Unseen facial and bodily expressions trigger fast emotional reactions. *Proceedings of the National Academy of Sciences* 106: 17661–6.

17. K. C. Berridge and P. Winkielman (2003) What is an unconscious emotion? The case for unconscious 'liking'. *Cognition and Emotion* 17: 181–211.

18. Ibid.

19. E. T. Rolls (2005) *Emotion Explained*. Oxford University Press, Oxford.

20. N. H. Frijda (1986) *The Emotions*. Cambridge University Press, Cambridge.

M. Mendl, O. P. Burman, and E. S. Paul (2010) An integrative and functional framework for the study of animal emotion and mood. *Proceedings of the Royal Society B-Biological Sciences* 277: 2895–904.

21. T. Halliday and H. P. A. Sweatman (1975) To breathe or not to breathe: the newt's problem. *Animal Behaviour* 24: 551–61.

22. N. B. Davies, M. D. L. Brooke, and A. Kacelnik (1996) Recognition errors and probability of parasitism determine whether reed warblers should accept or reject mimientic cuckoo eggs. *Proceedings of the Royal Society Series B.* 263: 925–31.

23. M. Mendl, O. H. P. Burman, R. M. Parker, and E. S. Paul (2009) Cognitive bias as an indicator of animal emotion and welfare: emerging evidence and underlying mechanisms. *Applied Animal Behaviour Science* 118: 161–81.

24. M. Bateson and S. M. Matheson (2007) Performance on a categorisation task suggests that removal of environmental enrichment induces 'pessimism' in captive European starlings (*Sturnus vulgaris*). *Animal Welfare* 16: Suppl. S33–S36.

25. See note 6.

26. J. Balcombe (2006) *Pleasurable Kingdom*, p. 27: 'gripped by a stifling "behaviorist" dogma that rejected the idea as conscious, feeling beings'.

   M. Bekoff (2007) *The Emotional Lives of Animals.* New World Library. Novato, CA.

27. D. M. Rosenthal (2005) *Consciousness and Mind.* Oxford University Press, Oxford.

   P. Carruthers (2004) Suffering without subjectivity. *Philosophical Studies* 121: 99–125.

28. R. W. Elwood, S. Barr, and L. Patterson (2009) Pain and stress in crustaceans? *Applied Animal Behaviour Science* 118: 128–36. P. P. G. Bateson (1991) Assessment of pain in animals. *Animal Behaviour* 42: 827–9.

29. H. Wuerbel (2009) Ethology applied to animal ethics. *Applied Animal Behaviour Science* 118: 118–27.

CHAPTER 7

1. H. Wuerbel (2009) Ethology applied to animal ethics. *Applied Animal Behaviour Science* 118: 118–27. Wuerbel makes a similar point. He argues that

we need to define animal welfare in terms of the 'integrity' of the functioning organisms rather than using anthropomorphic ideas of sentience and consciousness.

2. D. Fraser (2008) *Understanding Animal Welfare: The Science in its Cultural Context.* Universities Federation for Animal Welfare, Wiley-Blackwell, Chichester.

3. World Health Organization, www.who.int/zoonoses/vph/eu (accessed 10 November 2011).

4. H. J. Godfray, J. R. Beddington, I. R. Crute, L. Haddad, D. Lawrence, J. F. Muir, J. Pretty, S. Robinson, S. M. Thomas, and C. Toulmin (2010) Food security: the challenge of feeding 9 billion people. *Science* 327: 812–17.

5. Food Standards Agency survey. http:www.food.gov.uk/multimedia/pdfs/biannaalpublicattitudestrack/pdf/ (accessed 10 November 2011).

6. P. Rozin, C. Fischler et al. (1999) Attitudes towards large numbers of choices in the food domain: a cross-cultural study of 5 countries in Europe and the USA. *Appetite*: 46: 304–8.

7. National Opinion Poll (2010) www.farmsfoodandfuel.org/user (accessed 10 November 2011).

8. http://www.ichartsbusiness.com/channels/why-do-people-buy-organic-foods (accessed 10 November 2011).

9. R. E. Kahn, D. F. Clouser, and J. A. Richt (2009) Emerging infections: a tribute to the one medicine, one health concept. *Zoonoses and Public Health* 76: 66–73.

10. F. J. Hoerr (2010) Clinical aspects of immunosuppression in poultry. *Avian Disease* 54: 2–15.

11. T. Humphrey (2006) Are happy chickens safer chickens? Poultry welfare and disease susceptibility. *British Poultry Science* 47: 379–91.

12. P. D. Warris (1998) The welfare of slaughter pigs during transport. *Animal Welfare* 7: 365–81.

13. R. Bonney (2008) The business of farm animal welfare. In *The Future of Animal Farming: Renewing the Ancient Contract*, ed. M. S. Dawkins and R. Bonney. Blackwell, Oxford, pp. 63–72.

14. Ibid, para 2.

15. P. Singer, *The Expanding Circle: Ethics and Sociology*. Oxford University Press, Oxford.

16. E. Chivian and A. Bernstein (eds) (2008) *Sustaining Life: How Human Health Depends on Biodiversity*. Oxford University Press, Oxford.

17. H. Wuerbel, (2009) Ethology applied to animal ethics. *Applied Animal Behaviour Science*. 118: 118–27.

CHAPTER 8

1. G. G. Simpson (1983) *Fossils and the History of Life*. W. H. Freeman, New York and San Francisco.

2. M. S. Dawkins (2001) How can we recognize and assess good welfare? In *Coping with Challenge: Welfare in Animals Including Humans*, ed. D. M. Broom. Dahlem University Press.

3. D. M. Broom (1998) Welfare, stress and the evolution of feelings. *Advances in the Study of Behavior* 27: 371–403.

4. E. M. Blass and A. N. Epstein (1971) A lateral preoptic osmosensitive zone for thirst in the rat. *Journal of Comparative Physiology and Psychology* 76: 378–94.

5. P. Berthold (1993) *Bird Migration: A General Survey*. Oxford University Press, Oxford.

6. S. Jenni-Eiermann and L. Jenni (1996) Metabolic differences between the post breeding, mating and migratory periods in feeding and fasting passerine birds. *Functional Ecology* 10: 62–72.

7. J. Clutton-Brock (1981) *Domesticated Animals: From Early Times*. Heinemann, London.

8. S. R. Osbourne (1977) The free food contra-freeloading phenomenon: a review and analysis. *Animal Learning and Behavior* 5: 221–35.

9. Just as a matter of terminology here, some people refer to primary and secondary needs. Fred Toates calls primary needs those where, if they are not met, the animal dies. Food, water, and shelter are primary needs. Secondary needs are those that, if they are not met, the animal does not die, but may behave as though the doing of the behaviour were very important to it. Chasing balls might be an example. Secondary needs are also sometimes called 'ethological' or 'behavioural' needs, but these terms cause such endless confusion (P. Jenson and F. M. Toates (1993) Who needs behavioural needs? Motivational aspects of the needs of animals. *Applied Animal Behaviour Science* 37: 161–81) that it is probably best to reserve the term 'needs' for what Toates calls primary needs, defined by the very starkest of requirements—the animal dies if the need is not met. But that then leaves us with a gap in how to describe the other sort of need, where the animal appears to be able to survive perfectly well even if it is not met. F. Toates (2004) Cognition, motivation, emotion and action: a dynamic and vulnerable interdependence. *Applied Animal Behavior Science* 86: 173–204.

10. N. Tinbergen (1951) *The Study of Instinct*. Oxford University Press, Oxford.

11. M. Bateson et al. (2006) Cues of being watched enhance cooperation in a real-world setting. *Biology Letters* 2: 412–14.

12. M. S. Dawkins (2003) Behaviour as a tool in the assessment of animal welfare. *Zoology* 106: 383–7.

13. R. A. Hinde (1970) *Animal Behaviour*, 2nd edn. McGraw Hill, New York.

14. J. M. McNamara and A. I. Houston (2008) Optimal annual routines: behaviour in the context of physiology and ecology. *Philosophical Transactions of the Royal Society*, Series B 363: 301–19.

15. European Union Welfare Quality: 'It is now widely accepted that animal welfare is very complex, that it can be affected by many factors and that it embraces both physical and mental health', http://www.welfarequality .net (accessed 10 November 2011).

16. M. S. Dawkins (1980) *Animal Suffering: The Science of Animal Welfare*. Chapman and Hall, London; D. M. Broom (1988) Welfare, stress and the evolution of feelings. *Advances in the Study of Behavior* 27: 317–403; G. Mason and M. Mendl (1993) Why is there no simple way of measuring animal welfare?

Animal Welfare 2: 301–19; M. Mendl (2001) Animal husbandry: assessing the welfare state. *Nature* 197: 31–2.

17. Measures of welfare include: Play: S. D. E. Held and M. Spinka (2011) Animal play and animal welfare. *Animal Behaviour* (in press); fractals: K. M. D. Rutherford et al. (2004) Fractal analysis of animal behaviour as an indicator of animal welfare. *Animal Welfare* 13: S99–S103.

18. J. F. Hurnik (1993) Ethics and animal agriculture. *Journal of Agricultural and Environmental Ethics* 6 (Suppl.): 21–35; G.P. Moberg (ed.) (1985) *Animal Stress*. Animal Physiological Society, Bethesda, MD.

19. T. G. Knowles, S. C. Kestin, and S. M. Haslam (2008) Leg disorders in broiler chickens: prevalence, risk factors and prevention. Public Library of Science research article. *PLoS One* (Feb.), Issue 2 e. 1545.3(2): e1545.

20. R. Dantzer (2001) Cytokine-induced sickness behavior: where do we stand? *Brain, Behavior and Immunity* 15: 7–24.

21. J. Ladewig and D. Smidt (1989) Behavior, episodic secretion of cortisol and adrenocortical reactivity in bulls subjected to tethering. *Hormones and Behavior* 23: 344–60.

22. M. Kiley (1972) The vocalizations of ungulates, their causation and function. *Zeitschrift für Tierpsychologie* 31: 71–122; F. Wemelsfelder (2007) How animals communicate quality of life: the qualitative assessment of behaviour. *Animal Welfare* 16 (Suppl.): 25–31.

23. P. H. Zimmerman, P. Koene, and J. A. R. A. M. Van Hoof (2000) The vocal expression of feeding motivation and frustration in the domestic laying hen Gallus gallus domesticus. *Applied Animal Behaviour Science* 69: 265–73.

24. A. Boissy et al. (2007) Assessment of positive emotions in animals to improve their welfare. *Physiology and Behavior* 92: 375–97.

25. M. B. M. Bracke and H. Hopster (2006) Assessing the importance of natural behavior for animal welfare. *Journal of Agricultural and Environmental Ethics* 18: 77–89.

26. The Farm Animal Welfare Council (FAWC) lists the Five Freedoms as:

    1. Freedom from Hunger and Thirst—by ready access to fresh water and a diet to maintain full health and vigour.
    2. Freedom from Discomfort—by providing an appropriate environment including shelter and a comfortable resting area.
    3. Freedom from Pain, Injury or Disease—by prevention or rapid diagnosis and treatment.
    4. Freedom to Express Normal Behaviour—by providing sufficient space, proper facilities and company of the animal's own kind.
    5. Freedom from Fear and Distress—by ensuring conditions and treatment which avoid mental suffering: http://www.fawc.org.uk/freedoms.htm (accessed 10 November 2011).

27. H. Kruuk (1976) The biological function of gulls' attraction towards predators. *Animal Behaviour* 24: 146–53.

28. J. Webster (1994) *Animal Welfare: A Cool Eye Towards Eden*. Blackwell Science; Oxford.

29. D. Fraser (2008) *Understanding Animal Welfare: The Science in its Cultural Context*. Univerities Federation for Animal Welfare, Wiley-Blackwell, Chichester.

CHAPTER 9

1. D. Lack (1933) Habitat Selection in birds. *Journal of Animal Ecology* 2: 239–62.

2. G. Orians (1971) Ecological aspects of behaviour. In *Avian Biology*, vol. 1, 2: 513–46 ed. D. S. Farner and J. R. King. Academic Press, New York.

3. J. Wah, I. Douglas-Hamilton, and F. Vollrath (2006) Elephants avoid costly mountaineering. *Current Biology* 16: R527–R529.

4. R. Constantine, D. H. Brunton, and T. Dennis (2004) Dolphin-watching tour boats change bottlenose dolphin (*Tursiops truncatus*) behaviour. *Biological Conservation* 117: 299–307.

5. R. D. Swetnam, J. D. Wilson et al. (2005) Habitat selection by yellowhammers *Emberiza citrinella* on lowland farmland at two spatial scales: implication for conservation and management. *Journal of Applied Ecology* 42: 270–80.

6. J. A. Hogan (1989) The interaction of incubation and feeding in broody junglefowl hens. *Animal Behaviour* 38: 121–38.

7. G. L. Charlton, S. M. Rutter, M. East, and L.A. Sinclair (2011) Preference of dairy cows: indoor cubicle housing with access to a total mixed ration vs. access to pasture. *Applied Animal Behaviour Science* 130: 1–9.

8. H. A. V. de Weerd et al. (1998) *Applied Animal Behaviour Science* 55: 369–82.

9. S. Buijs, L. J. Keeling, and F. A. M. Tuyttens (2011) Using motivation to feed as a way to assess the importance of space for broiler chickens. *Animal Behaviour* 81: 145–51.

10. J. J. Cooper, L. McAfee, and H. Skinn (2002) Behavioural responses of domestic ducks to nipple drinkers, bell drinkers and water troughs. *British Poultry Science* 43: S17–S18.

11. T. A. Jones, C. D. Waitt, and M. S. Dawkins (2009) Water off a duck's back: showers and troughs match ponds for improving duck welfare. *Applied Animal Behaviour Science* 116: 52–7.

12. G. J. Mason, J. J. Cooper, and C. Clareborough (2002) Frustrations of fur farmed mink. *Nature* 410: 35–36.

13. For example: J. L. Volpato (2009) Challenges in assessing fish welfare. *Institute for Laboratory Animal Research Journal* 50: 329–37; C. E. Manser, H. Elliott, T. H. Morris, and D. M. Broom (1996) The use of a novel operant test to determine the strength of preference of flooring in laboratory rats. Institute for *Laboratory Animal Research* 30: 1–6.

14. L. A. Hanmer, P.M. Riddell & C. M. Williams (2010) Using a runway paradigm to assess the relative strength of rats' motivation for enrichment objects. *Behavior Research Methods* 42: 517–24.

E. G. Patterson-Kane, M. Hunt and D. Harper (2002) Rats demand social contact. *Animal Welfare* 11: 327–32.

L. Carbone (2004) *What Animals Want. Expertise and Advocacy in Laboratory Animal Welfare Policy.* Oxford University Press, Oxford.

15. Ibid.

16. J. Rushen (1986) Aversion of sheep to electro-immobilization and mechanical restraint. *Applied Animal Behaviour Science* 15: 315–24.

17. E. A. Pajor, J. Rushen, and A. M. B. de Passille (2003) Dairy cattle's choice of handling treatments in a Y-maze. *Applied Animal Behaviour Science* 80: 93–107.

18. E. T. Rolls (2005) *Emotion Explained*. Oxford University Press, Oxford.

19. Ibid.

20. J. Stevenson (1967) Reinforcing effects of chaffinch song. *Animal Behaviour* 15: 427–32.

21. T. I. Thompson (1963) Visual reinforcement in Siamese fighting fish. *Science* 141: 55–7.

22. G. S. Losey and L. Margules (1974) Cleaning symbiosis provides a positive reinforcer for fish. *Science* 184: 179–80.

23. W. M. S. Russell and R. L. Burch (1959) *The Principles of Humane Experimental Technique*. Methuen, London.

24. E. M. Scott, A. M. Nolan, J. Reid, and M. L. Wiseman-Orr (2007) Can we really measure animal quality of life? Methodologies for measuring quality of life in people and other animals. *Animal Welfare* 16 (Suppl.): 17–24.

25. D. S. Arey (1992) Straw and food as reinforcers for prepartal sows. *Applied Animal Behaviour Science* 33: 217–26.

26. J. E. Van Der Harst et al. (2003) Access to enriched housing is rewarding to rats as reflected by their anticipatory behaviour. *Animal Behaviour* 66: 493–504.

27. A. I. Sandem and B. O. Braastad (2005) Effects of cow–calf separation on visible eye white and behaviour in dairy cows. *Applied Animal Behaviour Science* 95: 233–9.

28. D. M. Weary and D. Fraser (1995) Signalling need: costly signals and animal welfare assessment. *Applied Animal Behaviour Science* 44: 159–69.

29. K. C. Berridge (1996) Food reward: brain substrates of liking and wanting. *Neuroscience and Biobehavioral Reviews* 20: 1–25.

30. F. Wemelsfelder et al. (2001) Assessing the 'whole animal': a free choice profiling approach. *Animal Behaviour* 62: 209–20.

31. M. S. Dawkins (2006) Through animal eyes: What behaviour tells us. *Applied Animal Behaviour Science*. 100: 4–10.

32. M. S. Dawkins (2008) The Science of animal suffering. *Ethology* 114: 937–45.

33. T. C. Danbury, C. A.Weeks et al. (2000) Self-selection of the analgesic drug carprofen by lame broiler chickens. *The Veterinary Record* 146: 307–11.

34. J. J. Villalba, F. D. Provenza, and R. Shaw (2006) Sheep self-medicate when challenged with illness-inducing foods. *Animal Behaviour* 71: 1131–9; F. C. Colpaert et al. (2001) Opiate self-administration as a measure of chronic nociception pain in arthritic rats. *Pain* 91: 33–45.

35. C. M. Sherwin and I. A. S. Olsson (2004) Housing conditions affect self-administration of anxiolytic by laboratory mice. *Animal Welfare* 13: 33–8.

36. Reference back to discussion

CHAPTER 10

1. J. Balcombe (2006) *Pleasurable Kingdom*, pp. 27, 105. M Bekoff (2007) The Emotional Lives of Animals. New World Library, Novato, CA .

2. D. M. Broom (2010) Cognitive ability and awareness in domestic animals and decisions about obligations to animals. *Applied Animal Behaviour Science* 126: 1–11s.

3. P. Carruthers (2004) Suffering without subjectivity. *Philosophical Studies* 121: 99–125.

P. Carruthers (2005) Why the question of consciousness may not matter very much. *Philosophical Psychology* 18: 83–102.

4. R. W. Elwood and M. Appel (2009) Pain experience in hermit crabs? *Animal Behaviour*, 77: 1243–6.

5. S. J. Shettleworth (2010) Clever animals and killjoy explanations in comparative psychology. *Trends in Cognitive Sciences* 14: 477–81.

   C. Heyes (2008) Beast machines? Questions of animal consciousness. In *Frontiers of Consciousness*, ed. L. Weiskrantz and M. Davies. Oxford University Press, Oxford, pp 259–74.

6. T. Grandin (2008) Hard work and sustained effort required to improve livestock handling and change in dairy practices. In *The Future of Animal Farming: Renewing the Ancient Contract*, ed. M. Dawkins and R. Bonney. Blackwell, Oxford, pp. 95–108.

7. I am indebted to FAI for this succinct way of summing up the issue: http://www.faifarms.co.uk (accessed 10 November 2011).

8. While the answers are likely to be different in different cases, it is encouraging to know that in at least one case, the free-range environment did not seem to provide a disease hazard. *Campylobacter* are a major cause of food poisoning in humans and the most likely source of infection is through eating chicken. This had led to the suggestion that chickens should not be allowed to roam outside to prevent them from picking up *Campylobacter* from the 'dirty' outside environment. However, by using genotyping techniques to identify individual strains of *Campylobacter*, it was shown that free-range chickens did not seem to be picking up *Campylobacter* from their outside environment: F. Colles et al. (2008) *Campylobacter* infection of broiler chickens in a free-range environment. *Environmental Microbiology* 10: 2042–50.

9. B.E. Rollin (2007) Cultural variation, animal welfare and telos *Animal Welfare*. 16: 129–33.

10. Cochrane Collaboration. http//:www.cochrane.org (accessed 10 November 2011).

11. J. Webster (2005) *Animal Welfare: Limping Towards Eden* Universities Federation for Animal Welfare, Potters Bar.

M. Midgley (2006) *Animals and why they Mattev:* University of George Press, Athens BA.

P. Singer (1976) *Animal Liberation:* A New Ethics for Our Treatment of Animals, Jonathan Cape, London. R. Ryder (1975) *Victims of Science.* Davis-Poynter London.

12. B. E. Rollin (1990) *The Unheeded Cry.* Oxford University Press, Oxford.

INDEX